T0133931

Simulation of Beef Cattle
Production Systems
and Its Use
in Economic Analysis

About the Book and Editors

Economic analysis of beef cattle production has been limited by incomplete descriptions of the production process under which analysis takes place. With the exception of feedlot finishing, beef cattle production is the process of growing cattle on forage. Both the animal and its forage possess attributes of inputs and outputs of production. The production of forage constitutes one production process, weight gain by the animal represents another, and reproduction by female animals is yet a third process. Cattle reproduction involves all three processes, each influencing the outcomes of the others. The processes are complex in and of themselves and are further complicated when analyzed simultaneously.

The contributors provide several comprehensive models of beef cattle production systems. Overviews of the Texas A&M herd simulation model and the Kentucky BEEF models are coupled with studies using the models in economic analysis. Other chapters examine the extension of neoclassical economic theory to dynamic production processes, a model of DNA and protein synthesis, and the prospects of extending complex simulation models of animal production systems to beef cattle producers.

Thomas H. Spreen is an associate professor in the Food and Resource Economics Department of the University of Florida. David H. Laughlin is an associate professor in the Department of Agricultural Economics at Mississippi State University.

Simulation of Beef Cattle Production Systems and Its Use in Economic Analysis

edited by Thomas H. Spreen
and David H. Laughlin

Routledge
Taylor & Francis Group

LONDON AND NEW YORK

Chapter V was prepared by U.S. government employees as part of their official duties and legally cannot be copyrighted.

First published 1986 by Westview Press, Inc.

Published 2019 by Routledge
52 Vanderbilt Avenue, New York, NY 10017
2 Park Square, Milton Park, Abingdon, Oxon OX14 4RN

Routledge is an imprint of the Taylor & Francis Group, an informa business

Copyright © 1986 Taylor & Francis

All rights reserved. No part of this book may be reprinted or reproduced or utilised in any form or by any electronic, mechanical, or other means, now known or hereafter invented, including photocopying and recording, or in any information storage or retrieval system, without permission in writing from the publishers.

Notice:
Product or corporate names may be trademarks or registered trademarks, and are used only for identification and explanation without intent to infringe.

Library of Congress Cataloging-in-Publication Data
Simulation of beef cattle production systems and its
  use in economic analysis.
  (Westview special studies in agriculture science
and policy)
  Includes index.
  1. Beef cattle--Simulation methods. 2. Beef cattle--
Economic aspects--Simulation methods. I. Spreen,
Thomas H. II. Laughlin, David H. III. Series.
SF207.S387 1986   636.2'13   86-4065

ISBN 13: 978-0-367-28729-0 (hbk)
ISBN 13: 978-0-367-30275-7 (pbk)

Contents

Preface

Economic analysis of beef cattle production has been limited by the inability to fully describe the underlying production process. Except for confined feeding of cattle, beef cattle production is the process of growing cattle who consume forages. The animal and the forage possess attributes of both factors and products of production. The production of forage constitutes one production process, animal growth is another production process, and reproduction by female animals is a third production process. Cattle production involves all three processes in such a manner that each influences the outcome of the other. Each process is itself complex and analysis is further complicated when all three are considered simultaneously.

If we are unable to adequately describe such basic relationships as stocking rate, weight gain, calving rate, and culling rate, then correct measurement of inputs and outputs is impossible. Economic analysis becomes a matter of guesswork and the resulting conclusions questionable.

Recent advances in understanding basic biological relationships coupled with advances in computer simulation have spurred the devlopment of mathematical models of beef cattle production systems. This volume is a collection of papers in which various aspects of simulation of beef cattle production systems are discussed. Several prominent researchers in beef cattle simulation model development have contributed papers which document their individual efforts. Presentation of biological models are augmented with papers written by agricultural economists who have utilized these models in economic analysis.

In the first paper by Forbes and Oltjen, a historical perspective of systems simulation of beef cattle production is presented. They document the fact that the last ten years has brought significant advances in modeling beef cattle production systems. In the next paper, Trapp and Walker argue that for complex agricultural systems, such as beef cattle production, it is necessary to replace the traditional single output production function with a complex simulation model. Economic analysis of such a system cannot be conducted through traditional marginal analysis, but more sophisticated techniques are applicable.

Denham and Spreen provide an introduction to single-animal growth simulation models. Two widely recognized models of animal growth are reviewed. These two models along with another less well-known model are used to analyze a typical ration. The analysis reveals that widely differing results are obtained from the models suggesting that a consensus has not been reached regarding mathematical modeling of animal growth.

The next trio of papers deal with the Kentucky beef-forage model. The first paper by Loewer and Smith provides and overview of the work at the University of Kentucky. The next paper deals more specifically with the forage component of the model. Parsch, Loewer, and Laughlin show how the Kentucky model can be used in economic analysis.

The Texas A&M beef cattle simulation model is presented in the paper by Cartwright and Doren. Sullivan and Cappella detail the use of the Texas A&M model in several studies conducted at various locations throughout the world. Bourden and Brinks present their adaptation of the Texas A&M model to Colorado. They have made several notable modifications of the model and present results from economic analyses conducted with their model.

Oltjen presents a model of animal growth based on the basic building blocks of protein and DNA synthesis. This approach may lead to significant improvements over existing models.

Laughlin and Walden discuss the potential for delivering sophisticated bioeconomic simulation models of beef cattle to ranchers and farmers. They relate the experiences of extension specialists at Mississippi State University and conclude that large-scale models such as those presented in this book will not be made directly accessible to ranchers and farmers.

Thompson and Davidson address the problem of administering multidisciplinary in land grant universities. They conclude that multidisciplinary research is necessary to address the complex problems faced by today's agriculture.

With the rapid decline in the cost of computing it is likely that more researchers will become involved in the development of large-scale simulation models of beef cattle growth and reproduction. The editors hope this volume will facilitate the work of others in this field and provide an entry point for those new to this area of endeavor.

There have been other successful modeling efforts which have not been covered in detail in this book. Their omission is not an oversight nor intended to be a denigration of those works. Our aim was to include those large-scale models which have been utilized by economists. There may exist other simulation models of beef cattle production which have not yet been "discovered" by economists and could play a useful role in economic analysis.

The editors gratefully acknowledge funding provided by the Farm Foundation, Oak Brook, Illinois. Additional funding and other support was provided by the Institute of Food and Agricultural Sciences at the University of Florida. The agricultural experiment stations of Mississippi, Oklahoma, and Alabama also provided funding. This work was coordinated through the southern regional research project, S-172. Greg Sullivan played an important role in conceptualizing the structure of this book. The editors also thank Steve Denham and David Notter for constructive comments. Toni Glover and Lavon Mikell have done an excellent job typing the manuscript, and Berenda Williams helped prepare the manuscript.

Thomas H. Spreen
David H. Laughlin

I

Historical Perspective of Biological Simulation with Special Reference to Beef-Forage Systems

T.D.A. Forbes and James W. Oltjen

INTRODUCTION

The practice of agriculture is the management of biological systems with the object of economically producing one or more products. A great many biological systems, or more correctly sub-systems, are involved in the overall system called agriculture. The complexity of agricultural systems is well illustrated by the diagrammatic representation of sheep production given by Spedding (1975). To study such a system in its entirety in a traditionally designed experiment would be impractical if not impossible. However, it is theoretically possible to build mathematical models that describe the system, and which can be manipulated to predict the outcome of changing one or more of the input variables. This practice of systems analysis requires a comprehensive and quantitative description of the system before a model can be built (Smith, 1982). Generally, and particularly for models of forage-livestock production systems, the lack of knowledge of the plant-animal interface (Minson, 1983) and the traditional division of research areas between plant and animal scientists has hampered model development (Dougherty, et al., 1985).

This paper deals briefly with the historical development of simulation models, the objectives of models, and the problem of model validation with special reference to forage-livestock systems. It does not attempt to provide a comprehensive or in-depth review of simulation modeling, but rather highlights those areas that the authors feel are of particular importance.

University of Kentucky, Lexington, Kentucky and Oklahoma State University, Stillwater, Oklahoma.

HISTORICAL DEVELOPMENT OF SIMULATION MODELS

Mathematical models are commonplace in agricultural research; the most commonly used are regression equations. Before the development of the high-speed computer simple mathematical models could be and were computed, but the process was tedious and time-consuming. More complex models were seldom attempted for the simple reason that an error made early in the process might not be detected until near the end of the analysis, and to recalculate a computation that had already taken several weeks was not pleasant to contemplate (Jeffers, 1978). The development of computers and the associated hardware, such as plotters, and software, such as statistical packages, created a surge of interest in mathematical techniques that previously had been impossible to use. Among the developments at this time that have had an influence on biological simulation were the development of linear programming, dynamic and stochastic simulation techniques and the development of general systems theory (Forrester, 1961; Naylor, et al., 1968; Van Dyne and Abramsky, 1975). The ability to repeat large numbers of complex computations speedily and accurately enabled mathematicians and biologists to deal with vast quantities of data and the interactions within the data for the first time. Thus, it was possible to not only conceptualize, but to initiate the mathematical modeling of agricultural systems for the first time.

In 1963 Arcus indicated the role simulation modeling might have in agriculture. Since then there have been a number of conferences and workshops on the role of modeling in agriculture (Dent and Anderson, 1971; Dalton, 1975; Penning de Vries, 1977; Hillyer, et al., 1981), and a very large number of models have been published. A literature search using the computer index Agricola (Dialog Information Services, Inc.) revealed that there has been a large increase in the number of publications that described the modeling of cattle production systems. Table 1 indicates the relative increase in modeling texts between the periods 1970-1978 and 1979-1984.

This comparison is by no means comprehensive, and may underestimate the number of texts actually published in each period. The table may be biased if modeling texts used key words different to those chosen here. If these possible biases are ignored it is quite clear that there has been a considerable increase in modeling activity in the last six years, (1979-1984) compared with the previous nine (1970-1978). In 1977 the Agricultural Research Service (ARS) Modeling Committee reported that two percent of ARS activities involved modeling. Of 27 "projects" involving modeling only one was in the livestock or veterinary sciences.

Table 1. Number of Publications per Year in the
Periods 1970-1978 and 1979-1984 which
include the Key Words, BEEF, MODEL,
COMPUTER, SIMULATION, and/or SYSTEM.

Key words	Publications per year 1970-1978	1979-1984
BEEF and MODEL	29.8	62.0
BEEF and COMPUTER	9.8	19.0
BEEF and SIMULATION	3.7	9.8
BEEF and SYSTEMS	118.6	167.5
BEEF and MODEL and SIMULATION	2.0	7.3
BEEF and MODEL and SYSTEM	4.9	11.8

The key words used are composited as follows: BEEF;
cattle, cows, calves, beef. MODEL; model, models,
modeling, modelling. COMPUTER; computer, computers.
SIMULATION; simulation, simulations. SYSTEM; system,
systems.

Four were in plant science. Table 1 indicates that in
excess of 600 texts have been published since 1970 that
include Beef and Model either in the title or in the
key words. However, the number of texts that have been
published that involve some aspect of grazing is very
much smaller. A total of 13 texts were published in
the period 1970-1978 which included the key words Beef,
Model and Grazing, while 19 texts with the same key
words were published in the period 1979-1984. Much of
the early work on grazing systems illustrated "the
paucity of quantitative data" (Brougham, 1983), though
this had been foreseen at a very early stage (Arcus,
1963). The reason for this has been pointed out in a
number of recent publications (Joandet and Cartwright,
1975; Minson, 1983; Dougherty, et al., 1985), namely
that not a great deal is known about the plant-animal
interface. The plant-animal interface may be defined
as the interactions that occur between grazed plants
and the grazing animal. They included the effects of
plant structure on the ingestive behavior, digestibility and intake of the grazing animal, as well as the
effects of grazing on subsequent regrowth of the plant.
In those few cases where the plant-animal interface has
been included in a model it has been based on a relatively simple set of assumptions frequently inconsistent with real-life situations (Joandet and Cartwright,
1975).

Another indication of the increased interest in modeling agricultural systems is the production of journals specifically aimed at modeling: e.g., Agricultural Systems, which was first published in 1975. This has allowed modelers greater freedom to publish than was possible previously in the more traditional agricultural journals. Reviews of the modeling literature may be found in the publications of Joandet and Cartwright (1975), Van Dyne and Abramsky (1975), Christian (1981) and Black et al. (1982). A more recent development has been the publication of reviews on specific types of models. Chudleigh and Cezar (1982) reviewed eight beef grazing systems models, while Wallach, et al. (1984) reviewed the sub-models dealing with maintenance energy requirements of grazing sheep taken from 14 models of sheep grazing systems.

Model development has not proceeded in any sort of logical, hierarchical order, as defined by Van Dyne and Abramsky (1975), but as might be expected from scientists, models have developed in a purely arbitrary manner in response to the needs of the scientists. There has not been a progression of model development from the modeling of sub-systems or components of systems to the development of models of larger systems. Chudleigh and Cezar (1982) commented on the amount of duplication that has occurred from this fragmented development and suggested that the time has come for a more cooperative approach in order to prevent modeling from becoming nothing more than an interesting postgraduate exercise. One reason for the lack of communication between modelers is the dynamic nature of modeling itself. By the time one version of a model has been published the current form may be quite different. Another problem has been the lack of proper documentation of the published model, a criticism made both by Van Dyne and Abramsky (1975) and Chudleigh and Cezar (1982).

MODEL OBJECTIVES

A frequent criticism of published models has concerned the lack of clear objectives of the model. Often the critics find that the objectives appear to have been written last (Van Dyne and Abramsky, 1975; Chudleigh and Cezar, 1982). Successful modeling, however, requires that the objectives be clearly stated initially. This step in modeling may be the most important because it suggests a starting point - forcing the researcher to define the systems inputs and outputs. It also helps set the level of aggregation at which the model variables are represented. Finally, the objectives should include an ending point. The model is valid when it predicts something with some level of confidence. Thus, the objectives will include necessary evaluation criteria.

TYPES OF MODELS

There are nunmerous texts that describe the various types of mathematical models that have been used in agricultural research (Dent and Anderson, 1971; Dalton, 1975; Baldwin, et al., 1977; Dent and Blackie, 1979; France and Thornley, 1984). The most widespread models are statistical, and include such well-known methods as analysis of variance and regression. For the purpose of systems simulation, however, more complex models are required. These models are, often, composed of a number of sophisticated algorithms and functions together with simple linear or quadratic equations. Models used in biological simulation are usually dynamic and deterministic (Dougherty, et al., 1985). Deterministic models may be described as models that make definite predictions for variables without any associated probability distribution (France and Thornley, 1984). Hence, a change in an independent variable is followed by a consistent change in the predicted variable. An alternative type of model is the stochastic model. Stochastic models differ from deterministic models in that one or more random variables are incorporated in the model with the consequence that the output of successive simulations vary even though the input data and constants are the same. The inclusion of the random variable has the effect of introducing a degree of variability into the system that attempts to mirror the variability found in biological systems. The output from repeated simulations can be tested statistically in the same way as replicated experimental data can be tested. The use of stochastic variables in models is increasing, though at present weather data are the variables most commonly treated stochastically due to the size and comprehensiveness of that particular data base. Dougherty, et al. (1985) point out that statistical descriptors of most model parameters are largely unavailable. At present a grazing systems model is being developed at the University of Kentucky using weather data to provide stochastic variables (Smith, et al., 1985). Christian (1981) suggests that using variables other than weather to provide stochastic elements in models is unlikely to lead to any marked improvements and that the use of stochastic variables increases the chance of confusion instead of clarification.
Among the hundreds of models that have been built to describe or predict biological systems in the livestock and forage area five main groups stand out.

1. Rumen models: The majority of rumen models have been written by Baldwin and co-workers (Baldwin and Koong, 1982), but other important contributers include Mertens and Ely (1979), Black, et al. (1981) and France, et al. (1982).
2. Feed intake models: (Bywater, 1984; Forbes, 1977).
3. Sheep grazing models: Most of the sheep grazing models have been written by Australian scientists reflecting the great importance of the sheep industry in that country. (Armstrong, 1971; Arnold and Campbell, 1972; Christian, et al., 1974; Vickery and Hedges, 1972; Freer, et al., 1970), but there has been at least one model of sheep grazing from the United Kingdom (Sibbald, et al., 1979).
4. Beef production models: (Joandet and Cartwright, 1975; Loewer, et al., 1981, Loewer, et al., 1983; Sanders and Cartwright, 1979; Siebert and Hunter, 1977; Sullivan, et al., 1981; Oltjen, 1983).
5. Crop growth models: (Ross, et al., 1972; Johnson and Thornley, 1983; Smith and Loewer, 1983).

The above review in no way is intended to provide a comprehensive list of the modeling literature.

Though models are being written that examine the interactions between the grazing animal and the forage (Forbes, et al., 1985; Sibbald, et al., 1979) they suffer from the lack of detailed knowledge of the plant-animal interface.

MODEL VALIDATION

Data with which to validate computer models and thereby gain confidence in the usefulness, are frequently difficult to obtain. This is particularly the case for plant-animal interface models. One of the main problems is the scarcity of data from large-scale, comprehensive experiments where both plant and animal variables were recorded together. Of the data that is available much of it has been obtained in the British Isles from perennial ryegrass (Lolium perenne) pastures (Forbes and Hodgson, 1985; Hodgson, 1982; Jamieson and Hodgson, 1979; Parsons, et al., 1983). Though it is hoped that the same relationships obtained from one pasture type will hold true on others, it may not be so. For example, it appears that the relationship between herbage height and bite size which is positive in temperate ryegrass pastures is negative on taller, tropical pastures (Chacon and Stobbs, 1976). As a consequence of the shortage of data much of the data used in models is little more than "the inspired guesses of a few experts" (Chudleigh and Cezar, 1982). Furthermore, model validation is often not carried out at all. Van Dyne and Abramsky (1975) suggest that some of the models reviewed by them used the same data both to construct the model and to validate it.

One of the difficulties as far as the modeling of grazing systems are concerned, especially where the plant-animal interface is examined, has been the difficulty in collecting the detailed data required. There is very little data available for such variables as intake per bite, grazing time, rate of biting, diet composition on a quantitative basis, rumen volume and rate of passage of digesta. Moreover, the data on the effect of the grazing animal on pasture regrowth is even scarcer. It has been suggested that the reason why so much effort has been spent on modeling rangeland systems is because of the difficulty in collecting data by conventional means (Dougherty, et al., 1985). It will, however, be necessary to carry out some experiments in order to validate such models.

CONCLUSION

Models of forage-livestock systems are dynamic in that the modelers are continually updating, improving and refining them. As our knowledge base expands it is probable that they will increase in size and complexity with more sub-models being added. However, it is likely that as the updated models are validated and as sensitivity analysis indicates which sub-models are unimportant, these models will contract in size and complexity (Dougherty, et al., 1985). Ultimately some models may become suitable for extension, and on-farm use, rather than remaining in the exclusive domain of a limited number of systems-oriented scientists.

8

REFERENCES

Arcus, P.L. 1963. "An Introduction to the Use of
Simulation in the Study of Grazing Management Problems." New Zealand Society of Animal Production.
23:159-163.

Armstrong, J.S. 1971. "Modelling a Grazing System."
Proceedings of the Ecological Society of Australia.
6:194-202.

Arnold, G.W., and N.A. Campbell. 1972. "A Model of a
Ley Farming System, with Particular Reference to a
Sub-Model for Animal Production." Proceedings of
the Australian Society of Animal Production.
9:23-30.

Baldwin, R.L., and L.J. Koong. 1982. "Mathematical
Modelling in Analyses of Ruminant Digestive Func-
tion: Philosophy, Methodology and Application."
In, Y. Ruckebusch and P. Thivend (Eds.) Digestive
Physiology and Metabolism in Ruminants. Proceed-
ings 5th International Symposium, pp. 251-268.
Clermont-Ferrand.

Baldwin, R.L., L.J. Koong, and M.J. Ulyatt. 1977.
"The Formation and Utilization of Fermentation End
Products." In, R.J.J. Clarke and T. Bauchop (Eds.)
Microbial Ecology of the Gut. pp. 347-391. Aca-
demic Press, London.

Black, J.L., G.J. Faichney, and L.E. Sinclair. 1982.
"Role of Computer Simulation in Overcoming Limita-
tions to Animal Production from Pastures." In,
J.B. Hacker (Ed.) Nutritional Limitations to Animal
Production from Pastures. pp. 473-493. Common-
wealth Agricultural Bureaux, Farnham Royal.

Black, J.L., D.E. Beever, G.J. Faichney, B.R. Howarth,
and C.N. McGraham. 1981. "Simulation of the
Effects of Rumen Function on the Flow of Nutrients
from the Stomach of Sheep: I. Description of a
Computer Program." Agricultural Systems. 6:195-
219.

Brougham, R.W. 1983. "Practical Livestock-Forage
Systems: Model to Manger." In, J.A. Smith and
V.W. Hays (Eds.) Proceedings of the XIV International Grassland Congress. Lexington, Kentucky,
pp. 48-53.

Bywater, A.C. 1984. "A Generalised Model of Feed Intake and Digestion in Lactating Cows." Agricultural Systems. 13:167-186.

Chacon, E., and T.H. Stobbs. 1976. "Influence of Progressive Defoliation of a Grass Sward on the Eating Behaviour of Cattle." Australian Journal of Agricultural Research. 27:709-727.

Christian, K.R., J.S. Armstrong, J.L. Davidson, J.R. Donnelly, and M. Freer. 1974. "A Model for Decision-Making in Grazing Management." Proceedings of the XII International Grassland Congress. Part I, pp. 126-130. Moscow, USSR.

Christian, K.R. 1981. "Simulation of Grazing Systems." In, F.H.W. Morley (Ed.) Grazing Animals. World Animal Science, Bl, pp. 361-377. Elsevier Scientific Publishing Company.

Chudleigh, P.D., and I.M. Cezar. 1982. "A Review of Bio-Economic Models of Beef Production Systems and Suggestions for Methodological Development." Agricultural Systems. 8:273-289.

Dalton, G.E. 1975. Study of Agricultural Systems. Applied Science Publishers Ltd., London.

Dent, J.B., and J.R. Anderson. 1971. Systems Analysis in Agricultural Management. Wiley, Sydney.

Dent, J.B., and M.V. Blackie. 1979. Systems Simulation in Agriculture. Applied Science Publishers, London.

Dougherty, C.T., N. Gay, O.J. Loewer, and E.M. Smith. 1985. "Overview of Modeling for Forage and Beef Production." In, V.H. Watson and C.M. Welles (Eds.) Simulation of Forage and Beef Production in the Southern Region. pp. 3-8. Southern Cooperative Series Bulletin 308.

Forbes, J.M. 1977. "Interrelation between Physical and Metabolic Control of Voluntary Food Intake in Fattening, Pregnant and Lactating Mature Sheep: A Model." Animal Production. 24:91-101.

Forbes, T.D.A., and J. Hodgson. 1985. "Comparative Studies of the Influence of Sward Conditions on the Ingestive Behavior of Cows and Sheep." Grass and Forage Science. 40:69-77.

10

Forbes, T.D.A., E.M. Smith, R.B. Razor, C.T. Dougherty, V.G. Allen, L.L. Erlinger, J.E. Moore, and F.M. Rouquette. 1985. "The Plant-Animal Interface." In, V.H. Watson and C.M. Welles (Eds.) Simulation of Forage and Beef Production in the Southern Region. pp. 95-116. Southern Cooperative Series Bulletin 308.

Forrester, J.W. 1961. Industrial Dynamics. MIT Press, Cambridge, Mass.

France, J., and J.H.M. Thornley. 1984. Mathematical Models in Agriculture. Buttersworth, London.

France, J., J.H.M. Thornley, and D.E. Beever. 1982. "A Mathematical Model of the Rumen." Journal of Agricultural Science. (Camb.) 99:343-353.

Freer, M., J.L. Davidson, J.S. Armstrong, and J.R. Donnelly. 1970. "Simulation of Summer Grazing." Proceedings of the XI International Grassland Congress. Surfers Paradise, Queensland, Australia, pp. 913-917.

Hillyer, G.M., C.T. Whittemore, and R.G. Gunn (Eds.). 1981. "Computers in Animal Production." British Society of Animal Production Occasional Publication 5. British Society of Animal Production, Thames Ditton.

Hodgson, J. 1981. "Variations in the Surface Characteristics of the Sward and the Short-Term Rate of Herbage Intake by Calves and Lambs." Grass and Forage Science. 36:49-57.

Jamieson, W.S., and J. Hodgson. 1979. "The Effects of Variation in Sward Characteristics upon the Ingestive Behaviour and Herbage Intake of Calves and Lambs under a Continuous Stocking Management." Grass and Forage Science. 34:273-282.

Jeffers, J.N.R. 1978. An Introduction to Systems Analysis: With Ecological Applications. University Park Press, Baltimore.

Joandet, G.E., and T.C. Cartwright. 1975. "Modeling Beef Production Systems." Journal of Animal Science. 41:1238-1246.

Johnson, I.R., and J.H.M. Thornley. 1983. "Vegetative Crop Growth Model Incorporating Leaf Area Expansion and Senescence, and Applied to Grass." Plant, Cell and Environment. 6:721-729.

Loewer, O.J., E.M. Smith, G. Benock, T.C. Bridges, L.G.
Welles, N. Gay, S. Burgess, L. Springate, and D.
Debertin. 1981. "A Simulation Model for Assessing
Alternative Strategies of Beef Production with
Land, Energy and Economic Constraints." Trans-
actions of the ASAE. Vol. 24: 164-173.

Loewer, O.J., E.M. Smith, K.L. Taul, L.W. Turner, and
N. Gay. 1983. "A Body Composition Model for Predicting Beef Animal Growth." Agricultural Systems.
10:245-250.

Mertens, D.R., and L.O. Ely. 1979. "A Dynamic Model
of Fibre Digestion and Passage in the Ruminant for
Evaluating Forage Quality." Journal of Animal
Science. 49:1085-1095.

Minson, D.J. 1983. "Forage Quality: Assessing the
Plant-Animal Complex." In, J.A. Smith and V.W.
Hays (Eds.). Proceedings of the XIV International
Grassland Congress, Lexington, Kentucky.

Naylor, T.H., J.L. Balinfy, D.S. Burdick, and C. Kong.
1968. Computer Simulation Techniques. John Wiley
and Sons, New York.

Oltjen, J.W. 1983. "A Model of Beef Cattle Growth and
Composition." Ph.D. Thesis, University of California, Davis.

Parsons, A.J., E.L. Leafe, B. Collet, P.D. Penning, and
J. Lewis. 1983. "The Physiology of Grass Produc-
tion Under Grazing. II. Photosynthesis, Crop
Growth and Animal Intake of Continuously Grazed
Swards." Journal of Applied Ecology. 20:127-139.

Penning de Vries, F.W.T. 1977. "Evaluation of Computer Simulation Models in Agriculture and Bio-
logy: Conclusions of a Workshop." Agricultural
Systems. 2:99-107.

Ross, P.J., E.F. Henzell, and D.R. Ross. 1972.
"Effects of Nitrogen and Light in Grass-Legume
Pastures: A Systems Analysis Approach." Journal
of Applied Ecology. 9:535-556.

Sanders, J.O. and T.C. Cartwright. 1979. "A General
Cattle Production Systems Model. 1. Structure of
the Model." Agricultural Systems. 4:289-309.

Sibbald, A.R., T.J. Maxwell and J. Eadie. 1979. "A
Conceptual Approach to the Modelling of Herbage
Intake by Hill Sheep." Agricultural Systems.
4:119-134.

12

Siebert, B.D., and R.A. Hunter. 1977. "Prediction of Herbage Intake and Liveweight Gain of Cattle Grazing Tropical Pastures from the Composition of the Diet." Agricultural Systems. 2:199-208.

Smith, E.M. 1982. "Systems Research Provides New Knowledge about Agriculture." In, M.G. Russel, R.J. Sauer and J.M. Barnes (Eds.) Enabling Interdisciplinary Research: Perspectives from Agriculture, Forestry and Home Economics. pp. 155-159. Miscellaneous Publication 19-1982, University of Minnesota.

Smith, E.M., and O.J. Loewer. 1983. "Mathematical-Logic to Simulate the Growth of Two Perennial Grasses." Transactions of ASAE. Vol. 26, pp. 878-883.

Smith, E.M., L.M. Tharel, M.A. Brown, G.W. Burton, C.T. Dougherty, S.L. Fales, V.H. Watson, and G.A. Pederson. 1985. "The Plant Growth Component." In, V.H. Watson and C.M. Welles (Eds.) Simulation of Forage and Beef Production in the Southern Region, pp. 9-36. Southern Cooperative Series Bulletin 308.

Spedding, C.R.W. 1975. "The Study of Agricultural Systems." In, G.E. Dalton (Ed.) Study of Agricultural Systems, pp. 3-19. Applied Science Publishers, London.

Sullivan, G.M., T.C. Cartwright, and D.E. Farris. 1981. "Simulation of Production Systems in East Africa by Use of Interfaced Forage and Cattle Models." Agricultural Systems. 7:245-265.

Van Dyne, G.M., and Z. Abramsky. 1975. "Agricultural Systems Models and Modelling: An Overview." In, G.E. Dalton (Ed.) Study of Agricultural Systems. pp. 23-106. Applied Science Publishers, London.

Vickery, P.J., and D.A. Hedges. 1972. "A Productivity Model of Improved Pasture Grazed by Merino Sheep." Proceedings of the Australian Society of Animal Production. 9:16-22.

Wallach, D., J.M. Elsen, and J.L. Charpenteau. 1984. "Maintenance Energy Requirements of Grazing Sheep: A Detailed Comparison Between Models." Agricultural Systems. 15:1-22.

II

Biological Simulation and Its Role in Economic Analysis

James N. Trapp and Odell L. Walker

The single equation, production function model is deeply entrenched in minds and methods of agricultural economists. Our purpose is to explore the proposition that traditional production economics should be expanded to explicitly encompass biophysical simulation modeling. Important conceptual and pedological advantages are afforded by abstractions such as $Y = f(X_1, \ldots X_n)$ and convenient mathematical forms such as:

$$Y = aX_1^{b_1} X_2^{b_2}$$

to represent the biophysical relationship. They allow an illusion of completeness and practicality in theoretical development and exposition. However, users are often lax in specifying underlying assumptions and describing a more realistic firm production environment. We believe that biophysical simulation modeling provides the needed realism and flexibility required in application of production theory at the farm firm level.

Production economics and other literature is rich in analyzing problems of estimating and using biophysical production functions (e.g. Heady and Dillon, 1961; Anderson, 1975; Bradford and Debertin, 1983) suggesting modifications of traditional techniques (e.g. Antle, 1983; Anderson, 1981; de Janvry, 1972), and outlining alternate models (e.g. Dillon, 1976; Konandreas and Anderson, 1982; Anderson et al., 1977; Chavas and Kliebenstein, 1983; Just et al., 1983; Smith et al., 1977 and 1978). Georgeescu-Roegen (1972) presented a

Department of Agricultural Economics, Oklahoma State University, Stillwater, Oklahoma.

critique along epistemological lines. Articles by Musser and Tew (1984) and Boggess (1984) provide a recent review of biophysical simulation use in the U.S. and evaluate potential applications and advantages and disadvantages of simulation.

We emphasize three major issues. The first is that strong (mostly) empirical reasons exist to abandon traditional production functions. Secondly, production economists need to feel that reasonable substitutes exist for estimating biophysical relationships and for the calculus of optimization which is the foundation of our "trade." Thus, we present and evaluate promising developments to that end. Thirdly, we suggest some ways of integrating biophysical models and management control functions.

PRODUCTION FUNCTIONS AND
TRADITIONAL PRODUCTION ECONOMICS

An early empirical estimation of a production function was made by Cobb and Douglas in 1928 (not in the references). They specified an aggregate production function to estimate the value of industrial production where the inputs were labor and capital. They introduced and used the now familiar Cobb-Douglas production function form of Log Y = Log a + b Log L + c Log C. From their work evolved an important foundation for agricultural production economics and Heady's text, "Economics of Agricultural Production and Resource Use" which was published in 1952. This text was followed in 1961 by Heady and Dillon's "Agricultural Production Functions." More recently texts including those by Doll and Orazem (1978) and by Beattie and Taylor (1985), have been published, but the "state of the art" in agricultural production theory and application really has not advanced much since Heady's first text was published. The Beattie and Taylor text does increase the mathematical rigor with which the theory of production is addressed and sets a proper stage in our opinion for discussion of modeling in the following statement:

"Mathematical specification of the production function can range from simple algebraic functions, Y = f(x), to highly complex "systems" of equations, such as a detailed model of corn plant growth and response to the nitrogen fertilization rate. The degree of mathematical complexity of the production function depends on the problem and the degree of accuracy desired." (p. 3)

Perhaps later they will add a chapter or two showing how such systems are used in decision analysis.

The firm decision problem is to maximize an objective function for which some of the arguments are suggested in equation (1) (Hertzler, 1983):

$$\text{Max } U = f\ (M_T,\ Y_{it}\ ,\ D_t,\ V_t,\ m_t,\ C_t,\ W_t,\ r) \qquad (1)$$

where

M_T = total wealth at T,

Y_{it} = amount (head, acres, \$ value) of commodity, or resource i in time period t,

D_t = debt at time t,

T = years in planning horizon,

V_t = a measure of risk in the business (e.g. variance of income or probability that the firm will not survive),

m_t = money income at time t

C_t = direct consumption at time t,

W_t = days of work per year by the operator,

t = a time subscript t = 1,...,T, and

r = a discount factor for time preference.

Management decisions and exogenous events determine Y, M, m, D and r, while W and C depend on managerial, owner and/or family choices. All variables may affect utility directly or indirectly through other variables, e.g. W may affect Y, M and C; Y affects M and m; and C affects M and m. The biological-physical relationship ("production function") enters through Y, and M, D, C, and V may constrain Y.

The management-owner-family group maximizes equation (1) subject to resource constraints and production responses. Given a problem with a single product, two inputs, no resource limitaions, perfect knowledge, a one period planning horizon, and for $U = f(\pi)$:

$$\pi = P_y\ f_y\ (X_1,\ X_2) - P_X\ X_1 - P_X\ X_2. \qquad (2)$$

If π (profit) is approximated by $C_t + M_{t+1} - M_t$ (consumption plus gain in net worth in year t), maximizing (2) is roughly consistent with (1). Production economics texts usually derive optimum quantities of Y, X_1, and X_2 and speculate about possible $f_y(X_1,\ X_2)$ relationships and their effects on optimum quantities. Traditional production economics theory also covers statistical production functions, for example, as treated by Heady and Dillon (1961). Multiple products, economies of size and scale, activity analysis (e.g.

linear programming), and simple examples of relaxing timelessness and perfect knowledge assumptions round out traditional treatment of production economics theory.

In apparent conflict with theoretical training, the farm management professional soon discovers the virtual impossibility of providing biological relationships needed in analyses by simply fitting functions to available experimental or cross sectional data. In fact examples of useful production functions are hard to find in the recent literature. Among the problems with trying to apply simple production function concepts and solve equation (2) are:

    a. The production function is more completely described as:

$$Y = f = (X_1, X_2, X_3 \ldots X_K / X_{K+1}, \ldots X_n : Z_1, \ldots Z_n), \quad (3)$$

        where

        $X_1 \ldots X_K$ = controllable (decision) variables;

        $X_{K+1} \ldots X_n$ = variables predetermined for the enterprise and/or the farm for the planning period;

        $Z_1 \ldots Z_n$ = uncontrollable and, perhaps, unknown inputs.

        Heady and Dillon (1961, p.145) alerted us to this specification early in production function analysis. Important interactions probably exist among and within X_i and Z_i. It is difficult to find data to fit (3). As a result, reality is usually ignored and simplistic statistical functions result.

    b. Timing of inputs within the production period is an important part of the production process but it is usually ignored in production functions. In reality production steps are recursive. Production is more like a recipe emphasizing process management, rather than strictly a formula which only prescribes ingredients.

    c. Inputs may be non-homogenous, partly because of timing. For example, through a production season, pasture declines in quality, animals on feed get fatter and heavier, and sunlight gets less intense.

    d. Technology, economic conditions, and objectives (decision rules) may change within as well as across production periods. Thus, the timelessness assumption leaves too many important managerial alternatives and concerns out of the problem.

e. Data used to measure production and inputs in traditional production function analysis is usually highly aggregated. For example, fertilizer and feed inputs are usually for the whole period as are capital items (e.g. machinery costs). Whether to consider a whole plant or animal rather than basic processes within the living organism is a more fundamental aggregation problem. For example, DNA accretion and protein synthesis and degradation can be used to simulate post-weaning growth patterns of mammals (Oltjen et al., 1983). That is a long disaggregation step from fitting body weight to intake of grain, supplement and forage. Even the DNA model involves aggregation. Oltjen et al. (1983) suggest that a lower level of aggregation (more basic processes) improves response prediction across species and production conditions, but that the choice of aggregation level depends on intended uses of the model.

As mentioned earlier, recent examples of empirical production functions are not abundant in the literature, although we can cite a few (Harman et al., 1983; Hoyt, 1984). Some others attempt to analyze stochastic as well as decision variables (de Janvry, 1972; Just et al., 1983). Later research in that vein combines modeling tools with stochastic production functions to estimate yield distributions (Smith et al., 1984).

Harmon et al. (1983) present the following yield response function for wheat.

$$Y = 26.966 + 4.387X - .075X^2 + .2698N \qquad (4)$$

where

X = irrigation water (inches) applied April 1 - June 15
N = percent of normal rainfall April 1 - June 15
Y = irrigated wheat yield in the Texas Panhandle (Tam 105 Variety)

How useful is equation (4) in decision making? The analyst would probably set N at its historical average to evaluate the intermediate run profitability of irrigated vs. dryland wheat or to decide whether to invest in an irrigation system. Implicit assumptions would be that growing time, distribution of rainfall, and other conditions are at normal values. Unfortunately, short run decisions (within the production season) would not be supported well in view of problems cited earlier. The farmer should water when he projects an economic need during the production period, within constraints of his soil and irrigation system capacity. Interactions among production and input variables must be considered each time he waters. For example, what are the implications of a full irrigation followed by a big Texas Panhandle, spring rain? What if he does not water and no rain comes?

APPROACHES TO HANDLING RESPONSE ESTIMATION PROBLEMS

Heady and Dillon (1961) anticipated a possible need to use systems of equations rather than a single equation model, however, empirical applications are very scarce. If interdependence exists among inputs and outputs to the extent that some variables are determined within the system rather than exogenously, a simultaneous equation system is required. For example, management interacts with concurrent weather and pest conditions to determine irrigation, fertilizer, tillage, and pest control inputs. The level used interacts to determine yield and quality levels and harvest timing.

How many agricultural economists assuming a single equation production function have approached an agronomist or animal scientist asking for inputs and outputs? The biological scientist scratches his head and says "well it depends." He would much rather give "if-then" answers than averages. In fact, it is hard for him to talk about average relationships because input decisions are not made in that manner. To the extent that a biophysical model represents reality and the perceptions of the biological scientist, it should improve cooperation among sciences.

In this section, we discuss possibilities for recognizing and handling time, risk, and the full planning-implementation-control process.

Time

Dillon (1976) cites four influences of time in production response.

a. Production depends on time. For example, the wheat plant needs growing time under suitable temperature and moisture conditions.

b. The capacity of fixed inputs to accommodate variable inputs is a function of time. Moisture intake rate by soils and ability to deliver water through an irrigation system are examples.

c. The time sequence or pattern of input injection or timing of harvest may influence production. Certainly, receiving all of an aver-

age annual spring rainfall a few days before
harvest doesn't help wheat yield.
d.   Inputs may carry over from or to other periods.

Dillon (1976) analyzes "best operating conditions"
for selected time dependent response situations including: (1) single injection of a factor, multiple harvest over the later part of the response period and
interaction between inputs and the harvest pattern and
(2) batch problems in feeding broilers for which the
profit maximizing feed mix changes with broiler growth
(and time), the production response period (time) is a
function of the ration fed, and space required per
batch changes with broiler size (time).   He also outlines models for livestock production on pasture and
crop production with fertilizer carryover.   In each
case treated by Dillon (1976), a multiple equation
model is necessary.

Risk

If concern for risk is real as suggested in equation (1), a simple response function such as equation
(4) is not very useful.   The optimal level of irrigation (X) is certainly not independent of natural rainfall and other uncontrolled variables.   The decision
maker needs probability distributions of yields (or
profits) for different levels of irrigation (and other
decision variables).   Then, a decision criterion
reflecting risk attitude is needed to choose the preferred distribution.   Risk decision rules are not the
focus of this paper but the literature offers many good
possibilities (Roumasset, 1976; Anderson et al., 1977).

Planning-Implementation-Control

Farm management literature has followed the lead
of production theory in using the production function
in a timeless, perfect knowledge, and homogenous products and inputs framework.   Most farm management texts
discuss acquisition of planning data from records,
experiment stations, extension agents and other farmers.   But students are still often perplexed when they
find enterprise and whole farm budgeting exercises that
require them to use judgment and select bits of data
from numerous sources.   Sometimes they learn to use
sensitivity analysis, particularly in linear programming, to test implications of data "inaccuracy."
The new farm management book by Boehlje and Eidman
(1984) takes a step beyond using general production
function concepts in planning.   The section they label

implementation includes traditional resource acquisition and management issues. However, the section called control covers some key problems in carrying out production plans. Some involve implementation of plans in an "if-then" mode which they call concurrent control. For example, in their Pig Production System Case, the plan calls for five females to be bred per week (p. 693). If conception rates lag, more than five females must be bred the next week and adjustments made later in pig weaning ages to make room in farrowing crates. Such concurrent adjustments are important in determining net returns. Clearly, records and concurrent and feedback control principles are keys in such an operation. Bywater (1985) also provides excellent suggestions on integrating control concepts in a dairy production system. The crop irrigation system discussed by Boehlje and Eidman (1984, p. 713) also emphasizes concurrent control and conveys more useful information than equation (4) by itself.

OPTIMIZATION IN PLANT AND ANIMAL MODELING

Problems which lead to modeling rather than production function fitting are apparent from preceding discussions: (1) decision makers usually interact with other variables during the production period; (2) data are scarce, particularly concerning effects of uncontrollable variables; (3) technology and fixed factors of production change across time and statistical production functions quickly become obsolete; (4) in general, production is not timeless and inputs are not homogenous, and (5) data tend to be produced in bits and pieces which do not suit statistical production function estimation.

Our contention is that biophysical simulation modeling and optimal control theory, when properly wedded, will become the "New Theory of Production Economics." We attempt to outline some considerations for using such a new approach in the remaining portion of this section. This outline will draw upon the work of Anderson et al., (1977), Musser and Tew (1984), Boggess (1984), Antle (1983), Chavas and Kliebenstein (1983), and others who have made suggestions.

Many economists readily accept the biological scientist's viewpoint that a production function is best represented by a system of equations or "model." Boggess (1984), in his review of Musser and Tew (1984), aptly points out that "fundamentally, all biophysical simulation does is generate the production response surface which is necessary for all empirical production economics research." But he also points out that the typical structure of biophysical simulation models is "mathematically less tractable than the neoclassical production function. As a result, simulation analysis coupled with various search algorithms are normally used to analyze decision alternatives, rather than (for) analytical derivations of the 'optimal' input level." Hence, while economists may have been quick to recognize the realism of biophysical simulation models for describing a production response surface, we have been slow to apply such models to decision making problems because we do not know how to "optimize" them. Even if optimization is possible with a search algorithm, the theoretical properties of the optimum are often unknown or lost in the process, leaving theoreticians uneasy about the solution.

Indeed the problem of optimizing simulation models has caused many economists to take the view expressed by Musser and Tew (1984) that simulation "does not propose to identify 'optimal' plans for firm managers. Rather it proposes to provide information which most likely has qualitative value for farm managers." In some ways this view likens the simulation model to a computerized budgeting process. Within this view lies a major frustration for economists in their attempt to use biophysical simulation models as management decision aids. This frustration stems from two aspects. First, the biophysical model is often so detailed and disorganized that general recommendations can not be made from it. Likewise, it often can not be directly used by any one manager unless it is modified for his situation. The problem of proper aggregation level is not new to production economics. In the past, single equation production models have been too aggregated. Biophysical simulation models now tend to be too disaggregated.

The second source of frustration with biophysical simulation is that the traditional calculus of maximization no longer provides a workable means to find a solution, nor, in general, does linear programming. That leaves many economists stripped of a substantial part of their normal wares, and thus very unsure about their exposure in using simulation models. Linear programming does provide some hope for a workable optimizing procedure. Indeed, linear programming contains a very crude biophysical simulation model. But the biophysical models and physical scientist's natural views of them do not fit the structure of linear programming well. Most biophysical models are envisioned and specified as a series of sequential, stochastic steps. Such a model structure is most easily programmed into a series of functional relations that have an infinite number of branching possibilities. Such models do not lend themselves well to linear programming.

Use of simulation models is not new to economists. Macro simulation models of the economy have been constructed since the 1950's. Until recently the use of these models in a "what-if" simulation mold was accepted. Forecasts were made by assuming expected conditions and seeing what the model indicated. Musser and Tew (1984) suggest that economists use simulation models with the hope that the information provided would aid in the qualitative decisions of managers. But more recently economists have begun to study the use of policy to direct the economy toward desired states or objectives. In this pursuit they have begun to use optimal control theory and optimization procedures. Hence, economists do have a background in simulation model construction and use that can be helpful in aiding physical scientists in the construction of biophysical models. However, the control function and need for an optimization procedure is much more pressing in the production economics field than it has been in policy and price analysis.

In summary, while economists can readily envision the descriptive value of biophysical models, they can not readily envision their optimization. Herein we believe lies the economist's problem in using these models and in aiding in their construction such that they can be applied to management decision problems. But perhaps all is not lost. Several pieces of work have recently been published at the frontier of developing optimal solution procedures for dynamic/simulation models. It is our opinion that these works point the direction for the future of quantitative production economics and provide hope that someday we will have a "theory of dynamic production" suitable to guide the economic application of biophysical simulation models.

SEQUENTIAL DECISION MAKING

Antle (1983) provides an informative departure point for our discussion of advances in dynamic production theory. Antle formulates the production model as a sequence of interlinked production steps. Each step can be defined as an individual production function whose output feeds forward as an input into the next production stage. The major concern of the article is with the proper econometric estimation procedure for these interrelated production functions. But in addressing this problem, Antle gives an excellent summary of the differences between single equation static production theory and dynamic/sequential decision procedures. Given the sequential/stochastic nature of most biophysical models his framework provides a natural theoretical bridge from static production theory to biophysical models and on into optimal control theory.

Antle summarizes the structure of his sequential decision model as follows, where production is divided into T stages. Output of firm i in stage t is Q_{it}, with an input vector X_{it}, and coefficient vector B_t, and production disturbances E_{it}. He describes the model as a sequence of "stage production functions" that can be written as follows:

$$Q_{i1} = f\ (X_{i1},\ B_1,\ E_1)$$
$$\vdots \quad \vdots \quad \vdots \quad \vdots \quad \vdots \quad \vdots \tag{5}$$
$$Q_{it} = f\ (Q_{i,t-1},\ X_{it},\ B_t,\ E_t)$$

$$t = 2,\ldots,T;\ i = 1,\ldots,N$$

Assuming the final product Q_{it} is sold in period T at price P_{it} and input prices are W_{it}, profit is:

$$\pi_{iT} = (P_{iT} \times Q_{iT}) - \sum_{t=1}^{T} W_{it} \times X_{it} \tag{6}$$

Antle differentiates static one period solution problems from dynamic multiperiod sequential solution problems by the information utilized by the dynamic decision maker. The information used pertains to three features of sequential solutions:

a. Sequential dependence of decisions: Decisions made earlier may affect those made later. If the producer takes this into account, then his optimal input choice in period 1 may depend on how it affects optimal inputs in period 2.

b. Information feedback: Information that becomes available during earlier stages may be utilized in subsequent decisions. Producers may use knowledge of the actual output of period 1, rather than original estimates of period 1 production to determine optimal inputs in period 2.

c. Anticipated revision: Decisions made earlier may be revised later as new information becomes available. If a producer knows that information about period 1 outcomes will become available in period 2, his choices in period 1 can be based on the conditional distribution of period 2 outcomes (conditional on period 1's results) rather than on the unconditional distribution of period 2 outcomes.

Thus a producer's decisions may be different
if he knows he can revise his plans later
rather than having to rely on his initial
expectations.

Antle's differentiation of sequential versus
static decisions points out the inherent interrelated-
ness of dynamicness and risk. Differentiations b and c
become irrelevant if perfect knowledge is assumed.
Under such an assumption Antle shows that open loop
optimal control procedures are appropriate. If both
sequential dependence (a) and information feedback (b)
are assumed, then open loop with feedback optimal control is appropriate. Finally, if sequential dependence
(a), information feedback (b), and anticipated revision
(c) are assumed, closed loop optimal control procedures
are appropriate.

If firms maximize expected returns, the ith firm's
objective is:

$$\text{Max } E \ (\pi_{iT}) \text{ subject to (5) and (6)} \tag{7}$$

This control problem is a "terminal period problem" and
is a specific case of a more general, multiperiod model
in which output is sold in each period rather than in
just the final period.

An immediate problem with the "terminal period
problem" in biophysical simulation is that the terminal
period is not always obvious, particularly with livestock. When should the product be harvested, replaced?
The dynamic nature of biophysical simulation models
raises not only the sequential decision aspects posed
by Antle, but also the question of traditional asset
replacement/investment and disinvestment theory.

MODELING DYNAMIC AGRICULTURAL PRODUCTION
RESPONSE IN A CONTINUOUS TIME FRAMEWORK

Chavas and Kliebenstein (1983) (C&K) and Chavas et
al. (1985) address the problem of simultaneous optimal
input use and optimal replacement policy. They consider the case of an animal on feed. The problem is to
determine the daily intake rate and mixture of feed,
and the date at which the animal should be sold and
another put on feed in its place. They recognize that
this problem has been identified previously by Dillon
(1976) and Fawcett (1973). Dillon (1976) proposed the
more traditional solution of making the production
function a function of time and total input used during
the response period. Fawcett (1973) argues in favor of
using a differential (difference) equation to char-

acterize the growth process. He sights several advantages. First, it allows for the use of nutritional
information in the specification of the biological
growth. Second, it appears to be better suited than
Dillon's approach for discussing the effect of changes
in input use over time. In addition, C&K add that the
differential equation approach to the modeling of
dynamic processes corresponds to the "state equation"
specification in modern control theory. By optimizing
the model using modern control tools, the conditions
for efficient production in a dynamic setting can be
derived. Such conditions extend traditional production
theory in several ways. First, they refine and extend
Dillon's treatment of response efficiency over time.
Second, they simultaneously treat optimal input use and
optimal replacement policy.

C&K recognize that the first question is how to
model animal growth, and that this question is separate
from the question of deriving the economic implications
for production efficiency given an objective function.
C&K define their model as follows, beginning with a
differential growth function.

$$\dot{X}_t = f(X_t, U_t, t) \tag{8}$$

where

$\dot{X}_t = \dfrac{dx}{dt}$, the growth rate of an animal at time t;

X_t = the current level of production or animal weight;

U_t = a vector of input quantities (feed, veterinary care, etc.);

t = time.

The value of an animal at any point in time is

$$\pi = \int_0^T (Q_t Y_t - R_t U_t) e^{-it} \, dt + P_T X_T \, e^{-iT} - I, \tag{9}$$

where

I = initial fixed cost of the animal, including its purchase at time $t=0$;

$Y_t = g(X_t, U_t, t)$ is the current product obtained at time t during the growth of the animal (e.g. milk, wool, calves, etc.) and sold at the competitive market price Q_t.

Q_t = price of Y_t at time t; this price presumably can take into account changing product form/ quality as the animal grows and changes in price itself over time;

R_t = unit cost of input U_t;

i = discount rate, e^{-it} being the discount factor for a continuous time model;

T = replacement time (or marketing time) of the animal;

P_T = unit price of X_T, implying $P_T X_T$ is the "salvage value" of the animal; this price presumably takes into account changing product.

The objective function is assumed to be:

$$\text{Max } F = \sum_{j=0}^{N-1} \pi e^{-ijT} \qquad (10)$$

where j denotes the jth animal, and N is the number of successive animals planned to be involved in the pro- duction process. In other words, the objective is to maximize the net present value of net returns obtained from the production of a string of N animals.

Necessary conditions for the optimization of the above objective function and model have been developed in optimal control literature (Bryson and Ho, 1975; Kamien and Schwartz, 1981). Review of this development is beyond the scope of this paper. In essence, the conditions derived are similar to the traditional con- ditions for optimal replacement and optimal input use. However, they are now specified in relation to the production function $g(X_t, U_t, t)$. Optimal use of U_t is now defined as a vector of day-to-day input streams, rather than as a single value for the sum of inputs used over the production period.

The work of Antle (1983) with sequential produc- tion functions can conceptually be linked to model of Chavas and Kliebenstein (1983). Antle's sequential production function for each stage can be thought of as the integral of C&K's function Y_t. In a discrete sense the difference between two sequential stage production functions would define the differential production function used by C&K.

Given that the optimizing conditions for C&K's model are known, one must find the vector U_t and replacement period T that satisfy the optimal conditions subject to the production function constraint. Solution methods for such a task have been lacking in the past if U_t was a vector of any length or matrix of any significant size.

MULTIPERIOD OPTIMIZATION PROCEDURES

A good discussion of the current state of the arts in multiperiod optimization is contained in the December 1982 issue of the Western Journal of Agricultural Economics. Articles by Burt, Zilberman and Talpaz are contained in the issue. The articles are papers presented by them at an invited paper session.

Burt contrasts dynamic programming to optimal control theory. He favors the use of dynamic programming, but grants that it often is not compatible with the most intuitive and direct way to structure a model. Such is the case with most biophysical models and dynamic programming. All three of the papers presented in this session readily concede that while much has been written in the literature about control theory, actual numerical solution of applications is a rarity because most of the literature has been theoretical and general. Stated otherwise, not enough is known about the implementation of optimal control theory, and algorithms to solve optimal control problems are not well developed nor widely understood.

Talpaz, however, believes applied solutions of multiperiod dynamic models may become increasingly feasible as advances made during the 1970's in non-linear constrained optimization are implemented. As use of these algorithm's becomes more widespread, he argues software will become available which standardizes their application and allows them to be used routinely with a broad class of general simulation models. He sights MINOS/AUGMENTED as developed by Murtagh and Saunders (1981) at Stanford University as an example. Such solution algorithms use numerical methods to iteratively search for the vector of control variables that optimize the objective function specified. If and when such algorithms become generally available it would appear that economists will have the capability to optimize biophysical simulation models and have a theoretical basis from which to assure themselves of the properties of the optimal solutions.

The existence of effective multiperiod optimization techniques will permit economists to participate in biophysical modeling efforts by contributing what they are best trained to do. That is to integrate technical knowledge with measures of management objective values in order to develop management systems and decision criteria that best fulfill management's objectives. We now turn to a consideration of a general modeling approach to do this.

INTEGRATION OF BIOPHYSICAL MODELS AND MANAGEMENT CONTROL FUNCTIONS

Anderson (1981) provides the basic ideas for a crop model in a paper concerned with incorporating meteorological services into crop planning. We added animal and economic models to Anderson's formulations and made managerial interaction explicit (Figure 1). Following Anderson, data and models can be linked by: (1) estimating structural parameters using the data and scientific knowledge (e.g. path BE_2M_3 in Figure 1), or (2) using historical and/or subjective data directly as in a Monte Carlo approach (e.g. path BD_2M_3). If the later approach is used, the weather, pest and soil models are not operational because the "information" is incorporated in historical data.

We might conclude that a full climate model is unlikely in our lifetime, and agree to model path WDM. However, a pest model is a possibility. In fact, sub-models for different pests probably are needed. We know that crop and animal models are a reality and modeling paths such as $WD_1M\ 2 \to BE_2M_3 \leftrightarrow AE_3M_4$ are reasonable. Animal and agronomic data include fertility and irrigation experiments, time series and farm records data, and basic plant growth theory and observation. The later basic knowledge might be used to build a model to be tested against farm and experimental observations. Anderson (1981) suggests that hand sketched curves with the correct variables considered may prove more useful than least square estimates of inadequate production functions. He also suggests a technique for overcoming sparse data problems (Anderson, 1973).

Interdependence among variables is recognized by climate, soil moisture and pest model inputs to the animal and crop growth models, crop growth model input to the soil moisture and pest models and interaction between animals and crops. Time periods might be hours, days, weeks or months. Managerial control activities can be exercised in advance (preliminary control) or concurrently through each submodel. For example, the manager can observe soil blowing or crusting and use appropriate tillage or other practices to protect soil moisture. Machinery capacity is provided by advance planning (preliminary control). The manager can interact concurrently through the pest model (e.g. count insects and decide whether to spray), the crop model, (e.g. decide when to irrigate) the animal model, (e.g. culling heavy when pasture is scarce) and the

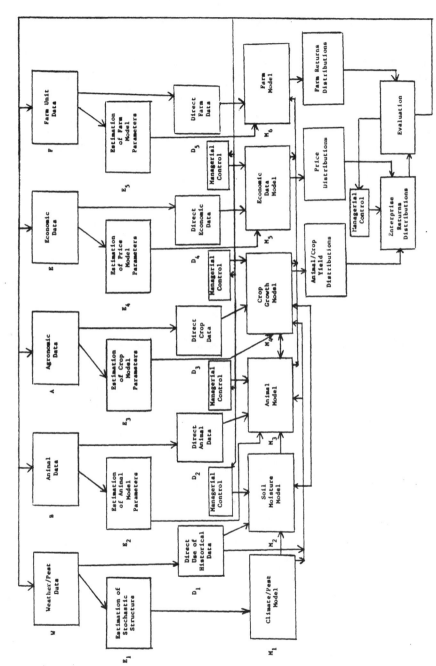

Figure 1. Modeling an Animal/Crop Enterprise and/or Farm Under Uncertainty

economic data model (e.g. price the inputs or crop by cash transactions, contracting or hedging).

If interest is only in the crop and animal enterprises, yield distributions can be produced through the model by generating uncontrollable variables from their distributions, setting variables to be held constant at selected levels and observing yield distributions for alternative variable input levels and practices. By adding price distributions for the inputs and products, distributions of enterprise returns are obtained. The evaluation step involves application of decision (choice) criteria, managerial learning, and feedback. Feedback may require new observation or experimentation to obtain new data.

Alternative modeling levels are readily illustrated by Figure 1. Estimation of animal model parameters in E_1, E_2, and E_3 may be at the process (e.g. protein synthesis) level. Animal and crop growth models must ultimately reach the animal level to allow enterprise evaluation. Of course, an enterprise usually involves multiple animals (a herd). The firm is the highest level of modeling included in Figure 1. Clearly, policy questions require aggregations of firm responses.

Economists may do process modeling in E_4 and E_5 but probably do not have much to offer in E_1, E_2, and E_3, except to communicate animal, enterprise and firm model needs. We suggest that managerial control provisions are needed beginning with animal and crop growth models and that optimization issues must be faced. If managerial control is included without optimization, the simulation model would be used in an "if-then" mode to generate information, as Musser and Tew (1984) seem to advocate. In either case (optimization or non-optimization), production economists could contribute by assuring that relevant managerial control actions can be accommodated in the biophysical model.

An advantage of ultimately working with a firm level model is that trade-offs among production, marketing, and financial decisions can be accommodated. For example, production risks may be offset by marketing and financing strategies elsewhere on the farm to increase attainment of objectives. Such realism is not possible in an enterprise model.

The model in Figure 1 is nearly impossible to optimize in a traditional analytical sense (e.g. directly through the calculus of optimizing a profit or utility function). Numerous and diverse (many non-normal) probability distributions and lack of data are major problems. Discrete decision events are another. However, the use of generalized non-linear constrained optimization procedures as suggested by Talpaz (1982) (such as MINOS/AUGMENTED) currently provide a mechanism for approximate optimization of such a system. Such procedures use heuristic search techniques to maximize a specified objective function. The truly generalized algorithm referred to by Talpaz functions in a manner such that no specific model structure is required. All that is needed is an operational simulation model that has control variables and state variables that provide usable measures of performance. In the model depicted in Figure 1, such an optimization procedure can be envisioned as a formalized "Evaluation Model" where output of the model is received and used to estimate "control functions" or "rules" to generate "optimizing" management controls.

Generalized optimization procedures proceed to experimentally determine approximate numerical values for the derivatives of a given objective function in terms of the selected control variables. The algorithm then directs its iterative search for a maximum based on these derivatives or gradients. Currently such algorithms are capable of considering up to about half a dozen control variables over a 20 to 30 year planning horizon. Talpaz (1982) is optimistic that these capabilities will be improved. Willingness to compromise the model's structure to fit the solution procedure will result in more powerful optimization potentials also, with the ultimate compromise being a return to structures compatible with linear programming or dynamic programming. Thus, currently a balance must be struck between model size and structure, and solution implementation.

SUMMARY AND IMPLICATIONS

We believe that agricultural economists don't approach production response questions with an appropriate mind-set and tool-set. Perhaps, the paucity of response studies in U.S. agricultural economics literature since the 1950's and 1960's reflects disallusionment and lack of success with production function results. Training in modeling concepts and approaches is needed. Professionals with such training should be able to work better with colleagues in other disciplines who appear to think in terms of systems, although they may not formally say so.

Plant and animal modeling is not without problems (Whelan et al., 1984). Complex models are hard to share (describe fully). An outsider must make a substantial investment in time and intellectual effort to learn important model details. However, the more experience one gains with models, the quicker key questions are raised and answered. A single equation production function transfers easily among users and use persists across time, particularly if users don't investigate the data base and statistical characteristics. Unfortunately, models seem to associate with originators. Only a few are widely adopted and applied over a period of years. Maybe models are too individualistic and each investigator thinks he/she can do a little better job. In any event the total cost of modeling is rather high and the cost per time used has historically also been high.

Modeling is expensive and time consuming. However, the microcomputer will probably cut expenses. Ideally models need to incorporate basic relationships which allow interpretation across variable levels, space and time. Models reveal knowledge gaps as well as provide forecasts.

In the agricultural production environment, full accuracy and optimality are only distant goals. Biophysical models use knowledge as it occurs in the real world -- deductive, tentative, piecemeal and short term validity. Subjects for models or sub-models need to deal with biological, physical and economic issues which make a real difference in decision choices.

REFERENCES

Antle, John M. 1983. "Sequential Decision Making in Production Models." American Journal of Agricultural Economics. 65:282-290.

Anderson, J.R. 1975. "One More or Less Cheer for Optimality." Journal of the Australian Institute of Agricultural Science, 41:195-197.

_____. 1981. Meteorological Services in Agronomic and Economic Evaluations of Risk, World Climate Program, proceedings of the Technical Conference on Climate - Asia and Western Pacific, WMO - No. 578, Geneva, pp 205-209.

_____. 1973. "Sparse Data, Climatic Variablity and Yield Uncertainty in Response Analysis." American Journal of Agricultural Economics, 55:77-82.

Anderson, Jock R., John L. Dillon, and J. Brian Hardaker. 1977. Agricultural Decision Analysis. Ames, Iowa: Iowa State University Press.

Anderson, W.K., R.C.G. Smith and J.R. McWilliam. 1977. "A Systems Approach to the Adaptation of Sunflower to New Environments, I. Phenology and Development." Department of Agronomy and Soil Science, University of New England, Armidale, N.S.W. (Australia), pp 141-152.

Beattie, Bruce and C. Robert Taylor. 1985. The Economics of Production. New York: John Wiley and Sons.

Boehlje, Michael D., and Vernon R. Eidman. 1984. Farm Management. New York: John Wiley and Sons.

Boggess, W.G., and C.B. Amerling. 1983. "A Bioeconomic Simulation Analysis of Irrigation Investments." Southern Journal of Agricultural Economics. 15:85-91.

Boggess, W.G., G.D. Lynne, J.W. Lores, and D.P. Swaney. 1983. "Risk-Return Assessment of Irrigation Decisions in Humid Regions." Southern Journal of Agricultural Economics. 15:135-143.

Boggess, William G. 1984. "Use of Biophysical Simulation in Production Economics." Southern Journal of Agricultural Economics. 16:87-90.

Bradford, Garnett L., and David L. Debertin. 1983. Establishing Linkages Between Production Theory and Enterprise Budgeting for Teaching and Extension Programs. Staff Paper #149, Department of Agricultural Economics, University of Kentucky, Lexington, KY.

Bryson, A.E., and Y. C. Ho. 1975. Applied Optimal Control. New York: John Wiley and Sons.

Burt, Oscar R. 1982 "Dynamic Programming: Has It's Day Arrived." Western Journal of Agricultural Economics. 7:381-394.

Bywater, A.C. 1980. "Development of Integrated Management Information System for Dairy Producers." Department of Animal Science, University of California, Davis.

Calkins, Peter H., and Dennis D. Di Pietre. 1983. "Successful Decisions in a Changing Environment." in Farm Business Management, New York: MacMillan Publishing Co., Inc..

Chavas, Jean-Paul, and James B. Kliebenstein. 1983. "On the Modeling of Dynamic Agriculturla Production Response." Contributed paper, American Agricultural Economics Association Annual Meetings, University of Illinois, Urbana, Illinois.

Chavas, Jean-Paul, James Kliebenstein, and Thomas D. Crenshaw. 1985. "Modeling Dynamic Agricultural Production Response: The Case of Swine Production." American Journal of Agricultural Economics. 67:636-646.

de Janvry, Alain. 1972. "Optimal Levels of Fertilization Under Risk: The Potential for Corn and Wheat Fertilization Under Alternative Price Policies in Argentina." American Journal of Agricultural Economics. 54:1-10.

Dent, J.B., and J.R. Anderson. 1971. Systems Analysis in Agricultural Management. New York: John Wiley & Sons.

Dillon, John L. 1976. The Analysis of Response in Crop and Livestock Production, second edition, University of New England, Armidale, Australia: Pergamon Press.

Doll, John P., and Frank Orazem. 1978. Production Economics, Theory with Applications. Columbus, Ohio: Grid Publishing, Inc..

Fawcett. R.H. 1973. "Toward a Dynamic Production Func-
    tion." Journal of Agricultural Economics. 24:543-
    555.

Georgescu-Roegen, Nicholas. 1972. "Process Analysis
    and the Neoclassical Theory of Production." Ameri-
    can Journal of Agricultural Economics. 54:279-294.

Harman, Wyatte L., John McNeill, and G.B. Thompson.
    1983. "Estimating Costs and Returns for Wheat Crop
    Alternatives in the Southern Plains--Problems and
    Data Needs", Technical Article No. 19033, Texas
    Agricultural Experiment Station, College Station,
    Texas.

Harris, T.R., and H.P. Mapp, Jr. 1980. "A Control
    Theory Approach to Optimal Irrigation Scheduling in
    the Oklahoma Panhandle." Southern Journal of Agri-
    cultural Econommics. 12:165-171.

Heady, Earl O. 1952. Economics of Agricultural Produc-
    tion and Resource Use. New York: Prentice-Hall.

Heady, Earl O., and John L. Dillon. 1961. Agricultural
    Production Functions. Ames, Iowa: Iowa State
    University Press.

Hertzler, Greg. 1983. "Dynamic Decisions for the Cow-
    Calf Producer." Staff Paper 132, Department of
    Economics, Iowa State University.

Hoyt, Paul G. 1984. Crop-Water Production Economics.
    Economic Implications for Colorado. Natural Re-
    source Economics Division, Economic Research Ser-
    vice, U.S. Department of Agriculture, Washington,
    D.C. ERS Staff Report NO. AGES 840427.

Just, Richard E., David Zilberman, and Eithan Hochman.
    1983. "Estimation of Multicrop Production Func-
    tions." American Journal of Agricultural Econom-
    ics. 65:770-781.

Kamien, M.I., and N.L. Schwartz. 1981. Dynamic Optimi-
    zation: The Calculus of Variation and Optimal
    Control in Economics and Management. Amsterdam:
    Elsevier North Holland, Inc..

Kliebenstein, James B., and Jean-Paul Chavas. 1983.
    "Toward a Dynamic Model of Swine Production Deci-
    sions." University of Wisconsin-Madison, College
    of Agricultural Economics.

Konandreas, Panos A., and Frank M. Anderson. 1982.
"Cattle Herd Dynamics: An Integer and Stochastic
Model for Evaluating Production Alternatives."
International Livestock Center for Africa, Addis
Ababa, Ethipoia, ILCA Research Report No. 2.

Konandreas, Panos A., Frank M. Anderson, and John C.M.
Trail. 1983. Economic Trade-Offs Between Milk and
Meat Production Under Various Supplementation Lev-
els in Botswana. ILCA Research Report, No. 20.

Murtagh, B.A., and M.A. Saunders. 1978. "Large Scale
Linearly Constrained Optimization." Mathematical
Programming. 14:41-72.

_____. 1981. "A Projected Lagrangean Algorithm
and Its Implementation for Sparse Nonlinear Constraints." Technical Report SOL-80-IR, Department
of Operations Research, Stanford University.

Musser, Wesley N., and Bernard V. Tew. 1984. "Use of
Biophysical Simulation in Production Economics."
Southern Journal of Agricultural Economics. 16:77-
86.

Oltjen, James W., Anthony C. Bywater, and R. Lee Baldwin. 1983. "Simulation of Normal Protein Accretion
in Rats." Journal of Nutrition. 115:45-52.

Roumasset, James A. 1976. "Measuring the Risk of In-
vesting in Nitrogen Fertilizer." Rice and Risk.
New York: American Elsevier Publishing Company,
Inc..

Smith, Joyotee, Gloria Umali, Mark W. Rosegrant, and
Abraham M. Mandc. 1984. Fertilizer and Risk in
Rainfed Bicol, Philippines. International Rice
Research Institute, Philippines.

Smith, R.C.G., S.D. English, and Hazel C. Harris. 1978.
"A Systems Approach to the Adaptation of Sunflower
to New Environments, IV, Yield Variability and
Optimum Cropping Strategies." Department of Agro-
nomy and Soil Science, University of New England,
Armidale, N.S.W.k (Australia).

_____. 1977. "A Systems Approach to the Adapta-
tion of Sunflower to New Environments, III, Yield
Predictions for Continental Australia." Department
of Agronomy and Soil Science, University of New
England, Armidale, N.S.W. (Australia).

Talpaz, Hovav. 1982. "Multiperiod Optimization: Dynamic Programming vs. Optimal Control." Western Journal of Agricultural Economics. 7:407-412.

Whelan, M.B., E.J.A. Spath, and F.H.W. Morley. 1984. "A Critique of the Texas A & M Model When Used to Simulate Beef Cattle Grazing Pasture." Agricultural Systems. 14:81-84.

Zilberman, David. 1982. "The Use and Potential of Optimal Control Models in Agricultural Economics." Western Journal of Agricultural Economics. 7:395-406.

III

Introduction to Simulation of Beef Cattle Production

S.C. Denham and T.H. Spreen

INTRODUCTION

Given a topic as broad as indicated by the title above, it seems logical to attempt a working definition of bioeconomic simulation as a starting point. Bioeconomic simulation is, to the authors, a mathematical description of an agricultural production system. The model derived is likely to be an empirical paradigm rather than be strictly obtained from a theoretical basis. The rationale behind such a choice is that the bioeconomic model is to be used as a management decision making tool rather than as a research model. As a result, there is generally greater concern for the quality of outputs than for the quality of the theory by which those outputs are calculated. Such concern for accuracy is necessary if the model is to be of any practical use to the beef producer, but the failure to appropriately apply theory often renders these models inadequate when dealing with production systems which differ greatly in climate, animal breed or body type, feedstuffs, and marketing situations.

Consider the highly aggregated model depicted in Figure 1. Given inputs such as the genetic potential of an animal, the environment in which that genetic potential may be expressed, and the feed resource utilized by the animal, a model should be capable of predicting such animal production characteristics as growth and/or reproduction and provide some accounting of animal retention or longevity. For several reasons, the feed resource is considered separately from the production environment. Primary among these reasons is

Department of Animal Science and Food and Resource Economics Department, University of Florida, Gainesville, Florida 32611.

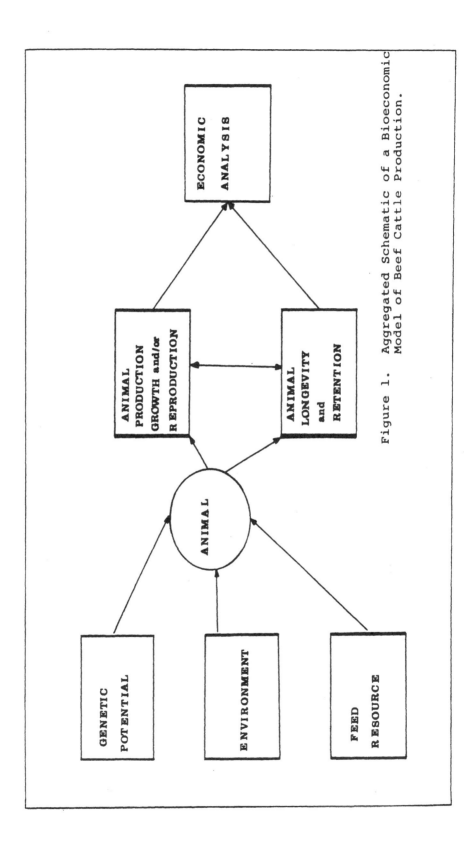

Figure 1. Aggregated Schematic of a Bioeconomic Model of Beef Cattle Production.

the more well defined linkage between animal performance and feeding level. Assuming the biological model is operating effectively, the outputs can then be utilized to make economic projections.

A major goal of animal scientists since the late 19th century has been to predict animal performance given a fixed feed resource. The converse problem, to predict the feed requirement necessary to support a fixed level of production, has been addressed over the same period. The development of feeding standards generally reflects this dichotomy. The many feeding systems in use throughout the world operate on this "fix one-predict the other" principle. Unfortunately, beef cattle cannot read any of these systems and so persist in operating as whole systems, utilizing feedback and feed-forward regulation to feed, grow, reproduce, and die in an integrated manner. If, as modelers, we fail to recognize this integrated system, then the models we develop will lack the ability to respond appropriately and adequately to a wide range of inputs.

In particular, consider how the values presented in the varying feeding systems are obtained. A fixed ration, or rations containing graded levels of the feedstuff in which we are interested, is fed to a group of animals under tightly controlled environmental conditions. If energy values of the feedstuff are being determined, then intake is fixed by the investigator and changes in body composition are either measured by slaughter balance or inferred by calorimetric measurement. Such problems as associative effects between feedstuffs and relative efficiency of energy utilization at various intake levels must be dealt with by assuming linearity of response. As a result, the values obtained should be used only in a similar context if we expect to accurately predict performance from intake or vice versa. Moving to a less controlled situation, we should expect poorer predictability.

Contrast, then, the ability of modelers to accurately forecast animal growth in a feedlot with its relatively fixed environment and feed source with our current ability to predict cow-calf performance on grasslands. Consider just the feed resource. Existing successful range models generally assume a predominately monoculture range resource of uniform quality and digestibility. While this may closely approximate the conditions in tame pastures and crop aftermath production systems, it is certainly inappropriate for extensive ranges with a multitude of plant species varying in composition botanically, geographically, and temporally. Further, range cows are not in the habit of eating at fixed levels. The intake of forage is dynamically intertwined with production level. Even if a specific level of production is fixed, the possible manifold of vectors describing intake on range is quite large. When we further consider how forage quality changes over time as a function of grazing behavior, it becomes all too tempting for the modeler to throw up his hands in disgust, shouting "What's the use of continuing?"

How have successful models proceeded? Most have tried to link intake and performance in a manner which adapts the feeding standards approach to the environment and feed resource for the site to be modeled. Availability and quality of feedstuffs are updated periodically within the framework of the model. These factors determine potential intake. Feed consumed then generates a potentially utilizable nutrient source which in combination with body reserves is used in a hierarchical arrangement. Radical changes in environment and the concomitant changes in the feed resource necessitate recalculation of many model parameters. Can both feed intake and animal performance be predicted with accuracy and precision? It is necessary to do so if further calculations of economically important variables such as stocking rate, investment in pasture improvement, and marketing of product are to be carried out. A short review of the most widely used systems for calculating animal growth and voluntary feed intake (VFI) follows in an attempt to demonstrate the strengths and weaknesses of each as a component of bioeconomic models of beef production.

MODELS BASED ON THE NET ENERGY SYSTEM

The net energy system has been widely adopted in the United States as the basis for growth simulation models. It is based on the now classic article by Lofgreen and Garrett (1968) who analyzed data on English breed implanted steers and heifers fed high energy diets in the Imperial Valley of California. At that time, the notion that intake should be partioned into feed for maintenance and feed for gain, and that feed for maintenance is utilized more efficiently than feed for gain, represented a marked improvement over the TDN system.

If performance is fixed, e.g. in a growing calf if gain is fixed, the net energy system provides a direct method to compute feed requirements.

(1) $NE_m = .077W^{.75}$

(2) $NE_g = .0557\ G^{1.097}\ W^{.75}$ for medium frame steers,

(3) $NE_g = .0686\ G^{1.119}\ W^{.75}$ for medium frame heifers,

where

NE_m = Mcal of net energy required for maintenance,

NE_g = Mcal of net energy required for gain,

G = daily live weight gain (kg),

W = current live weight (kg).

There is hierarchical arrangement between NE_m and NE_g in that feed is assumed to be first used to meet the maintenance requirement, and intake beyond that required to meet maintenance is said to be available for gain. The Nutrient Requirements of Beef Cattle (NRC) (National Academy of Sciences, 1984) includes an extensive list of feeds and forages found in the United States and for each feed gives estimated NE_m and NE_g concentrations (Mcal/kg). For example, consider a ration which is 80 percent corn and 20 percent corn silage (dry matter basis). The NE_m concentration of this ration is 2.066 Mcal and the NE_g concentration is 1.294 Mcal. A 400 kg medium frame steer is to gain 1 kg per day. Its daily NE_m requirement is 6.887 Mcal. Thus 6.887/2.066 = 3.333 kg of the ration is required to meet the maintenance requirement. The NE_g requirement for 1 kg/day gain is 4.982 Mcal. Thus 4.982/1.294 = 3.850 kg of the ration is required if the animal is to gain 1 kg/day. Total required intake is 7.183 kg per day. By confronting the animal with several rations which vary in their NE_m and NE_g concentrations and cost per unit, least cost rations can be determined via linear programming.

An economist may be interested in the problem given an animal and a feed source of known quality, what weight gain can be expected? Fox and Black (1977, 1984) have modified the net energy system to address this question. A simplified representation of the gain prediction model is

(4) $VFI = (.1493[NE_m] - .0460[NE_m]^2 - .0196)W^{.75}$

(5) $NE_m = .077W^{.75}$

(6) $NE_g = (VFI - NE_m/[NE_m]) \ [NE_g]$

(7) $G = 13.91 \ NE_g^{0.9116} \ W^{-0.6837}$ for medium frame steers,

(8) $G = 10.96 \ NE_g^{0.8936} \ W^{-0.6702}$ for medium frame heifers,

where

VFI = daily dry matter intake (kg),

$[NE_m]$ = net energy for maintenance concentration of the feed (Mcal/kg),

$[NE_g]$ = net energy for gain concentration of the feed (Mcal/kg),

and W, NE_m, NE_g, and G are as before.

In this model, dry matter intake depends upon the net energy for maintenance concentration of the feed and current animal weight. The feed required for maintenance is subtracted from total intake leaving the feed available for gain. This residual is converted to net energy for gain, which is used to predict gain. The gain equations (7) and (8) are the inverted forms of (2) and (3). For growing animals, this model has been generalized by Fox and Black (1984) to account for differences due to frame size, growth implants, and feed additives. It has been adapted to forage based systems by Brorsen, et al., (1983) and Spreen et al. (1985).

INTAKE

Ad libitum VFI varies due to several factors. Among these are age of cattle (Holmes et al., 1961) and energy density and/or crude fiber content of feed (Montgomery and Baumgardt, 1965; Nelson et al., 1968). The latter idea is best covered by Baumgardt (1970) and may be summarized to state that both physical and metabolic inputs interact to regulate VFI. The physical input may best be described as a distention mechanism, operating such that high fiber - low digestibility rations inhibit intake due to increased rumen fill. The metabolic input can be treated as feed intake controlled by physiological demand for energy and represents an upper limit to VFI beyond which the animal must be force-fed to increase intake. Mathematical treatment of these two inputs has been done for beef cattle (Song and Dinkel, 1978), sheep (Forbes, 1977a; Black, 1984), and dairy cattle (Forbes, 1977b).

However, the two major feeding systems for beef cattle in place today - referred to as the NRC (1984) and the ARC (1980) - derive intake equations quite differently. The equations are reflective of the basis of the energy requirement embodied in the feeding system. The NRC system bases intake prediction on metabolic body size and the net energy for maintenance concentration in the ration fed (equation 4). The ARC system, bases intake on metabolic body size, metabolizability of the dietary energy, and concentrate proportion. Further separate equations exist for coarse and fine diets (equations (9) and (10), respectively).

(9) $VFI = W^{.75}(.1065\ q + .037\ p + .0241)$,

(10) $VFI = W^{.75}(-.0466\ q + .1168)$,

where VFI and W are as above, q is diet metabolizability, and p is decimal proportion of concentrate included in the diet. The concept of metabolizability is the keystone of the ARC system. The ratio of metabolizable energy concentration [ME] to gross energy concentration [GE] of a feedstuff is defined to be the metabolizability of a feedstuff or ration. This value is relatively constant for a given diet.

The NRC system has additional correction for frame size and sex not found in the ARC system. On the other hand, the ARC does accomodate the differences in inputs to the system by providing separate equations for diets which would tend to first satisfy a physical limitation to appetite (equation 9) and for diets which would first satisfy metabolic requirements (equation 10). Still, both suffer from the drawback pointed out by Roux and Meissner (1984):

> "...The ARC and the NRC systems are based on cross-sectional analyses for which each individual is measured only once, during age or mass intervals convenient for experimentation. Inference to other stages of growth is generally by extrapolation based on tangential information."

This need not be taken as fully damning as might appear at first. In many situations, we wish to predict performance only for a short period, say a week, and then recalculate at the end of that time. This step-wise iterative approach is commonly used in several bioeconomic models currently available (Fox and Black, 1977; Spreen et al., 1985; Long et al., 1975). It has the advantage in these models of being concise programmatically. The models work well when describing performance in systems similar to those which have substantially the same data base. Transport of the models - that is, applying them to substantially differing inputs such as forage types and species differences - has not been spectacularly successful in most cases.

As an extension of the NRC system, the model of Fox and Black (1977, 1984) is probably the most widely known. This is a full scale bioeconomic model, front-ending the biological portion onto an extensive economic results package. The model adjusts feed intake through consideration of environmental factors relating to animal housing, inclusion of the effect of non-nutritive feed adjuvants and implants, and the impact of dairy breeding.

These factors are accomodated by using the "equiv-alent weight" concept. By use of conversion factors, animals being simulated are "converted" to their equiv-alent weight, which corresponds to animals similar to those upon which the NRC requirements were originally based. This method proved so successful that many of the current NRC requirements now incorporate this prin-ciple. The model is satisfactory for feedlot rations (high concentrate) and is used extensively in the field where available. Adequate data as to the performance of the model when confronted with such widely divergent inputs as all-forage rations or high Zebu percentage animals is not currently available.

Recent work by Spreen et al. (1985) has adapted the Fox and Black framework to production systems which are forage based. The model uses the concept of Forage Quality Index proposed by Moore (1978) and quantified by Golding et al. (1976). Forage Quality Index is an index of the voluntary intake of a particular forage. A Quality Index factor of 1.0 means that the voluntary intake of forage just meets maintenance requirements. Moore et al. (1981) have estimated Forage Quality Index for several tropical grasses.

If supplemental feeds are included in the diet, the Forage Quality Index is used to estimate a substi-tution factor which gives the reduction in forage intake per unit increase in supplemental intake. High Forage Quality Index implies nearly complete one-to-one substitution of forage for supplement while a low Qual-ity Index factor means near additivity.

The model has been used successfully for systems utilizing high quality small grains forages, but has been less accurate for lower quality tropical perennial summer grasses.

A different approach, growing out of the ARC metabolizable energy requirements, has been put forward by Roux and Meissner (1984). Using an allometric-autoregressive method, much of the correlation between successive measurements of VFI and growth can be removed. The biological basis for VFI prediction lies predominantly in assessing the flow dynamics of the rumen. This strongly addresses the physical input to VFI. The metabolic input is not accomodated directly, and the method would probably overpredict intake of very high energy diets. Additionally, the parameters estimated by this model have been shown to be time (and hence feed) invariant. In fact, most of the system can be reworked to make cumulative feed intake the indepen-dent variable. This derivation, however, goes beyond the scope of this paper. Therefore, long-term predic-tion of animal performance is possible without updating the input to the model. Changes in diet necessitate recalculation, and this may present some problems for prediction of animal performance under range conditions, where changes in forage populations and quality may occur. This presents a major difficulty for all the models covered, however, and can be partially ameliorated by breaking the time span considered into smaller intervals during which a relatively constant diet can be selected.

Other methods of predicting VFI have been developed, which are more dynamic in nature. Song and Dinkel (1978) iteratively calculate VFI for both physical and metabolic input based on considerations of digestibility of the organic matter in the diet and the rate of passage and excretion. While considerable accuracy is obtained, the model is not concise programmatically and requires estimates of digestibility and rate of excretion of each dietary component. As mixed rations differ substantially from their components with respect to these parameters, calculation of VFI by this model is somewhat constrained. Inclusion as a subroutine in mainframe type models with a rapid cycle time is probably the most appropriate use for this model.

The models proposed by Forbes (1977a, 1977b) are more integrated in nature. A full description here would be impossible, but the two inputs are fully operant and are regulated in turn by estimates of gut fill (for physical limitation) and maintenance plus specified rate of tissue accretion (for metabolic limitation). The circularity of growth dependent upon intake dependent upon growth ad infinitum is avoided by targeting growth rate as an input. This has a drawback in production oriented models. If the user has prior knowledge of performance on a diet, why bother predicting that performance? The equational forms, however, are strongly based on biological fact and these models stand up well to a variety of inputs.

Black (1984) has presented a model for sheep which predicts both VFI and animal growth. This simulation model, working primarily on a metabolizable energy basis for energy partitioning, includes an extensive submodel of rumen function. Diet is characterized chemically and physically, thus avoiding the problem of additive effects encountered when only the energy content of the feedstuffs is considered. Additionally, this allows for inclusion of protein effects on digestibility of the ration. The data presented by Black (1984) show that the model does an adequate job in predicting several performance variables. The author's stated drawbacks primarily center on use of empirical relationships rather than theoretical concepts in portions of the model and on the dependence of prediction upon adequate evaluation of the diet selected by grazing animals. The major limitations seen here are in redesigning the model for beef animals and its large size programmatically. As such, it does not provide a compact front-end model to evaluate alternative economic strategies without considerable cost to the user.

Several alternative models for predicting intake have been presented. It is instructive to compare these alternative methods. Table 1 presents estimates of dry matter intake for medium frame size steers (800 kg limit mass under the system of Roux and Meissner (1984)) at 100 kg intervals. The diet has a 5 to 1 forage to concentrate ratio and has a digestible energy concentration of 2.63 Mcal/kg dry matter. This composition corresponds to a metabolizability of .5 for the ARC system, a metabolizable energy concentration of 2.16 Mcal/kg dry matter, and a net energy for maintenance concentration of 1.30 Mcal/kg dry matter. The NRC and ARC fine diets are very close in predicted VFI. The ARC coarse diet predicts slightly lower intakes reflecting the input to fill obtained from feeding long forage. The method of Roux and Meissner (1984) is substantially different in predicted VFI, predicting greater intakes at moderate weights than either NRC or ARC, and much lower intakes as the animal approaches its limit mass. This reflects the conceptual difference between point estimation in the former two cases and the longitudinal estimate in the latter case based on cumulative feed intake as the independent variable rather than body weight.

Of all these models, that of Black (1984) seems best able to address VFI prediction across a wide variety of potential diets. Increases in microcomputer speed and memory may soon render the problem of large program size more tractable. The primary feeding standards and the models derived from them cannot currently predict VFI accurately for low quality diets of high roughage, and in particular the effects of protein supplementation on intake. Because of the inclusion of metabolizability of diets in the ARC system, protein effects are more likely to be accommodated accurately. However, this would require determination of q, diet metabolizability, across a wide spectrum of possible diets. A more theoretically based model should handle this in a better fashion.

GROWTH

Prediction of growth is rather more difficult than that of feed intake. First, there must be agreement on what exactly is being predicted. Live weight (LW) and empty body (EB) gains are most common. Other methods predict fat gain, muscle gain, protein deposition or other variables. From a biological point of view, prediction of empty body gain (EBG) seems most desirable, due to the extreme variability in fill from animal to animal. From an economic point of view, however, live weight gain (LWG) is a more important variable, as the marketplace trades in live animals, not empty bod-

Table 1. Predicted Ad Lib Feed Intake at 100 kg Weight Increments for Medium Framed Steers (800 kg Limit Mass) Fed a Diet Containing 2.63 Mcal/kg of Digestible Energy (1:5 Ratio Concentrate:Forage).

Live Mass (kg)	NRC[a]	ARC[b]		Roux and Meissner[c]
		Fine	Coarse	
100	3.1[d]	3.0	2.6	2.8
200	5.1	5.0	4.4	6.2
300	7.0	6.7	6.1	9.0
400	8.6	8.4	7.5	10.4
500	10.2	9.9	8.9	10.4
600	---	11.3	10.2	8.8

[a]Intake determined using predictive equations to calculate [NE_m] of given diet (1.297 Mcal/Kg DM).

[b]Values assume a metabolizability (g) of .5; fine diet assumes ground forage, coarse diet assumes long forage.

[c]Values determined from predicted ME intake ad lib feed intake and calculated ME of diet.

[d]All values carry units of kg/day.

ies. We again deal with the major systems in place, plus some others of interest.

No mention of growth can be made without addressing the concept of maintenance. The definition used by Armstrong and Blaxter (1984),

"...that level of energy required to be supplied daily in the feed in order that there may be zero change in the energy of bodyweight; thus, in terms of metabolizable energy (ME) it is the amount of ME per day that exactly equates to the daily heat production (HP) of the animal"

is a rigorous biological definition, and is used here, rather than defining maintenance as that level of energy required for either zero EBG or zero LWG, which would be somewhat more desirable from a financial viewpoint. This definition is one of the motivating forces behind the ARC feeding system. Efficiency of energy use can be reliably calculated at or near maintenance. Unfortunately, such calculation is difficult in the growing animal. Armstrong and Blaxter (1984) address this problem in some detail. They conclude that if maintenance is indeed fixed by this definition, then the efficiency of ME use for growth (k_f) must be viewed as variable and dependent upon levels of intake and diet composition. Given the data of Meissner and Roux (1984), it must also be considered to be a function of breed, previous level of nutrition, and body frame. Adequate accommodation of all these variables has not been made by either the NRC or ARC. Perhaps a better explanation for the curvilinearity of retained energy (RE) (net energy for growth, NE_g) is the increased energetic cost charged to metabolically active tissues at high ME intake. In some ways, this constitutes an increased maintenance requirement. With this hypothesis, only small changes in k_f need exist to yield the RE response to ME intake, and these changes may be attributed to variation in the end-products of digestion at the higher level of intake. Some evidence for changes in maintenance requirements may be found in Canas et al. (1982) and Ferrell et al. (1983).

In either case, prediction of growth is in large part dependent upon prediction of the composition of gain. Fat is accumulated more efficiently than protein due to a much lower rate of degradation and resynthesis (Armstrong, 1969), despite the fact that protein deposition should be somewhat more efficient than fat deposition on a theoretical basis (.84 for protein vs. .72 for fat, Baldwin, personal communication). An ideal system does not yet exist to accomodate this discrepancy, as predicting energy requirements per unit gain remains an inexact science.

The NRC predicts growth on a LWG basis in its tables, and presents equations for both LWG and EBG (equations 7 and 8). The basis for prediction is the NE_g content of the components of the diet, which are assumed to be additive. To facilitate comparison, however, NRC NE_g requirements are converted to ME, using .415 as an efficiency of ME use for growth. This is equivalent to a metabolizability (q) of .52 for the ARC system, which presents LWG predictions based on compositional data for EBG (actual EBG is not predicted explictly). However, by rearranging equations in Appendix 3.11 of ARC (1980), the following equation for LWG can be obtained:

$$G = \frac{Z \cdot k_f \cdot (1-\exp(-k_m \cdot MEI/Z))}{(4.1+.0332W-.000009W^2)+.1475 \; Z \cdot k_f \cdot (1-\exp(-k_m \cdot MEI/Z))}$$

where

$$k_m = \text{efficiency of ME use for maintenance}$$
$$MEI = \text{metabolizable energy intake (MJ, where } 1MJ = 4.182 \text{ Mcal)}$$
$$Z = \text{minimal metabolism}$$
$$= .0043W + .53(W/1.08)^{.67}$$

and W, G and k_f are as before, and the polynomial in W carries units MJ·d/kg as does the coefficient .1475. Suitable adjustment is made for sex and frame size. Table 2 presents ME requirements for gains of 1 kg per day for steers of three frame sizes at 100 kg increments as predicted by NRC, ARC, and Roux and Meissner (1984). While frame sizes and limit masses are not exactly convertible, the NRC requirements are substantially greater than either of the other two systems', especially at higher weights. Part of this may be attributed to differences in maintenance requirement estimates, but a difference in efficiency of ME use is also a large factor. The ARC system predicts a 16 percent improvement in ME use as compared to the NRC system for steers gaining 1 kg per day, while the system of Roux and Meissner (1984) projects a 20 percent greater efficiency. For steers at lower growth rates (Table 3), the improvements at 0.5 kg per day and 0.75 kg per day are 15 and 8 percent, respectively, for the ARC system and 56 and 29 percent, respectively, for the Roux and Meissner system. Disparities of this magnitude are far too great to be attributed to differences in maintenance requirements. It seems, therefore, that utilization of any of these systems for predicting growth in a bioeconomic model must be thoroughly justi-

Table 2. Estimated Metabolizable Energy Requirements (Mcal/Day) for Steers Growing at 1 kg/Day on a Diet Containing 2.63 Mcal/kg of Digestible Energy.

System	Frame size	Live mass (kg)					
		100	200	300	400	500	600
NRC	Small	10.6	17.8	24.2	30.0	35.5	---
	Medium	9.1	15.3	20.7	25.7	30.4	---
	Large	8.5	14.3	19.4	24.1	38.5	32.6
ARC	Small	9.1	13.4	17.4	21.0	24.4	27.5
	Medium	8.4	12.2	16.0	19.1	22.5	25.3
	Large	7.4	11.2	14.6	17.7	20.6	23.2
Roux and Meissner	525[a]	7.6	13.6	18.9	23.9	28.7	---
	800	7.6	12.7	17.2	21.3	25.1	28.7
	1400	8.1	12.7	16.2	19.6	22.7	25.6
	2000	8.6	13.1	16.7	19.8	22.5	25.1

[a] Limit mass in kg.

Table 3. Estimated Metabolizable Energy Requirements (Mcal/Day) at Different Growth Rates for Steers Fed a Diet Containing 2.63 Mcal/kg of Digestible Energy.

Live mass (kg)	NRC[a] Rate of gain (kg/day)			ARC[b] Rate of gain (kg/day)			Roux and Meissner[c] Rate of gain (kg/day)		
	0.5	0.75	1.0	0.5	0.75	1.0	0.5	0.75	1.0
100	6.7	7.4	9.1	6.0	7.4	9.1	4.1	5.7	7.6
200	10.4	11.8	15.3	9.1	11.0	13.4	6.7	9.3	12.7
300	13.6	15.7	20.7	12.0	14.3	17.4	8.8	12.2	17.2
400	16.7	19.2	25.7	14.3	17.4	21.0	11.0	14.8	21.3
500	19.6	22.3	30.4	16.7	20.1	24.4	12.7	17.0	25.1
600	---	---	---	18.9	22.7	27.5	14.6	19.4	28.7

[a]For medium frame size.
[b]For small frame size.
[c]For limit mass = 800 kg.

fied. As the NRC data base and equations are derived from predominately concentrate fed animals, it is not too surprising that the predicted efficiency of ME use on high forage diets is underestimated, since the prediction is near the extremes of the regression where a much larger error of prediction can be expected. A second source of error may be over-adjustment for size due to differences in mature body size of the same breeds on different continents.

Table 4 presents predicted growth rates for the three systems previously discussed, with ad lib intake calcuated on the basis of Roux and Meissner (1984). At this level of intake and energy density, the NRC predicted growth is substantially less than the ARC (66 percent) and the allometric-autoregressive (80 percent) systems. Roux and Meissner (1984) attributed the underprediction by the NRC, in its 1976 version, to a failure to accommodate frame size. The current version, even allowing for frame, differs little in its predictions relative to the ME based systems. This again reflects a lower efficiency of ME use for gain in the NRC system. In any case, it seems that empirical models such as these must be applied carefully to avoid inappropriately using the model outside its data base.

One large failing in these three models is a failure to incorporate the effect of dietary protein on gains. As essentially static models, all three convert dietary protein to body mass protein at a fixed rate. Growth is limited by restrictions on dietary protein below the amount required for stated gain, assuming a fixed ratio of protein to fat in the gain. Despite several attempts to circumvent the problem, the issue of protein to energy ratio in the diet is not adequately addressed. Changes in the composition of gain, which, according to Preston (1985), are not calculated properly in the NRC system, undoubtedly occur in a dynamic fashion. An adequate accounting of this effect cannot be reached on a long-term basis, and the effort involved in recalculating on a shorter term basis may reduce the ability to face this problem. Only Fox and Black (1977) and Gill et al. (1985) seem to have faced this problem, but there are still some problems with protein withdrawal after previous supplementation in the Fox and Black model.

Models of growth which are less empirical are covered elsewhere in this volume, but three closely related models of particular elegance should be mentioned. The work of Black (1984), Baldwin and Black (1979), and Oltjen et al. (1983) are notable. All are highly disaggregated and depend upon DNA/protein ratios to estimate lean tissue accretion. After the energetic cost of lean growth is accommodated, the remaining energy surplus/deficit is applied to fat deposition/ mobilization. These models represent a bold step

Table 4. Patterns of Growth Predicted by Different Systems for Medium Framed Steers (800 kg Limit Mass) Fed a Diet Containing 2.63 Mcal/kg of Digestible Energy, Based on ME Intake Predicted by the System of Roux and Meissner.

| Live mass (kg) | Predicted growth (kg/day) at | | | | | |
| | Ad lib intake | | | 70% of ad lib intake | | |
	NRC	ARC	Roux and Meissner	NRC	ARC	Roux and Meissner
100	0.42	0.50	0.80	0.05	0.19	0.61
200	0.77	1.00	1.06	0.33	0.59	0.81
300	0.97	1.13	1.13	0.40	0.70	0.86
400	0.77	1.08	1.06	0.32	0.57	0.81
500	0.54	0.89	0.90	0.20	0.40	0.69
600	0.20	0.66	0.66	---	0.00	0.50

toward a more accurate simulation of the growth process.

OTHER FACETS OF THE BEEF PRODUCTION SYSTEM

Some other productive functions in the cow-calf system which impact upon VFI and growth should be mentioned, although the primary objective in this paper is to address these two points. The first function of any brood cow, after staying alive, is the production of a live calf. The nutritional-reproductive interface has been an area of intense research at least since 1950. The work of Ferrell et al. (1976a, 1976b) is probably the primary base upon which the NRC bases its feeding recommendations. Application of dairy based data by Moe et al. (1972) also has been used to predict requirements for lactating animals. However, the objections raised in the introduction as to applying fixed intake methodologies to the ad libitum VFI-forage selection system on which beef cattle seem to operate apply here as well. Further, changes in maintenance requirements due to lactation and gestation need to be included. Finally, the effect of nutritional status on post-partum interval, with special attention to energy, nitrogen and phosphorus balance (and supplementation), should be more clearly addressed than is evident from published results.

Because most current models are set up for simulations under relatively fixed environmental conditions, the interaction between genotype and environment can usually be absorbed into the genetic component of the model. Specification of genetic potential is generally based on frame size (and proportion of dairy breeding for the model of Fox et al. (1977, 1984)) and milking ability. This is incomplete when dealing with production systems located in the southeastern United States. In this region, the proportion of Bos indicus breeding plays an important role in determining productivity. While existing models can accommodate this effect by judicious adjustment of system parameters, the adjustment is generally a rule-of-thumb type which makes comparison of animal types under different management systems much more difficult. The primary question to be answered with regard to conversion of feed to animal product is where the genotype-environment interaction is operating in the system. Possible points include behavioral changes related to intake, differences in digestive ability resulting in changes in the products absorbed into the animal's system, differential efficiency in use of these digestive end products, and changes in the hierarchical partitioning of metabolism. The latter two possibilities carry with them a change in the maintenance requirement of the animal, while the former two relate more closely to changes in VFI. Any or all of these may be operating and represent potential for error in prediction of performance.

INTERFACING BIOLOGICAL MODELS WITH ECONOMIC ANALYSIS

Despite the existence of several alternative models of intake and growth of beef cattle, there have been only rudimentary attempts to incorporate these models into economic analysis. Probably the most extensive effort has been by Fox et al. (1981), which is a microcomputer adaptation of the model of Fox and Black (1977). A similar model for stocker cattle growth available on a microcomputer has been developed by Brorsen et al. (1983). Using alternative intake equations, Spreen et al. (1985) has also developed a fully interactive computerized growth simulation model for pasture-based systems.

There have also been efforts to interface larger whole-herd models into economic models. Several efforts have been made with the Texas A&M model (e.g., Stokes et al., 1981), which are discussed elsewhere in this volume. Similar attempts with the Kentucky beef-forage model and the Colorado State model are also discussed elsewhere in this volume.

All of these efforts follow a similar mode of analysis. A production system is identified and the biological model simulates the system based upon a set of input data which is assumed to adequately describe the particular system of interest. Next the cost of the production system is estimated. This usually involves simply summing the components that must be purchased such as feed, fertilizer, labor, etc. Next the output of the system is valued. This usually involves "selling" the animals at market prices yielding an estimated gross return to the production system. Depending upon the economic questions to be answered, total cost, total return, or net return is reported.

In summary, the order of analysis is to first perform the biological simulation and second compute the economic implications of the simulation. In reality, there is feedback from economic performance to implementation of some management decision to biological performance of the animals. For example, consider the strategy of culling all open cows. This should increase the cash flow of the cattle operation and, over time, it should gradually improve the reproductive performance of the herd. Increased cash flow and increased reproductive performance suggest other management alternatives, namely improved pasture maintenance or retention of weaned calves to be sold at heavier weights. Capturing these potential feedback effects is beyond all known existing models.

SUMMARY

Some substantial differences and shortcomings of existing models for animal growth have been covered. These have been pointed out before and gradually progress is being made towards reconciling the differences. For scientists interested in simulating beef cattle production systems, the thrust of the argument presented here is that use of an empirical system to analyze a system that is outside the data base upon which it was constructed will probably give poor results. Empirical models will most likely require adjustment to a given production system. While this dashes hope that a single model of beef production can simulate any given system, it should give hope to those who are constructing theoretically based models that their efforts are more likely to be broadly applicable. However, until an adequate system of accurately predicting feed intake, given a growing animal consuming a constantly changing diet, is developed, the industry must rely upon the best available data to make projections. This means that the many available bioeconomic models available today each have a place and use in the beef cattle industry.

REFERENCES

Agricultural Research Council. 1980. The Nutrient
    Requirements of Ruminant Livestock. Commonwealth
    Agricultural Bureaux. The Gresham Press, Surrey.

Armstrong, G.D. 1969. In Handbuch der Tiernahrung,
    Vol. 1. W. Lenkeit, K. Breirem and E. Crasemann
    (Eds.) p. 385, Paul Parey, Hamburg.

Armstrong, D.G., and K.L. Blaxter. 1984. "Maintenance
    Requirement: Implications for its Use in Feed Eval-
    uation Systems." In Herbivore Nutrition in the
    Subtropics and Tropics. F.M.C. Gilchrist and R.I.
    Mackie (Eds.), p. 631. The Science Press, Craig-
    hall, R.S.A.

Baldwin, R.L., and J.L. Black. 1979. "Simulation of
    the Effects of Nutritional and Physiological Status
    on the Growth of Mammalian Tissue: Description and
    Evaluation of a Computer Program." Anim. Res. Lab.
    Tech. Paper No. 6, p. 1, CSRO, Australia.

Baumgardt, B.R. 1970. "Control of Feed Intake and
    Energy Balance." In Physiology of Digestion and
    Metabolism in the Ruminant. A.T. Phillipson (Ed.),
    p. 226. Oriel Press, Newcastle-upon-Tyne.

Black, J.L. 1984. "The Integration of Data for Pre-
    diction of Feed Intake, Nutrient Requirements and
    Animal Performance." In Herbivore Nutrition in the
    Subtropics and Tropics. F.M.C. Gilchrist and R.I.
    Mackie (Eds.), p. 648, The Science Press, Craig-
    hall, RSA.

Brorsen, B.W., O.L. Walker. G.W. Horn, and T.R. Nelson.
    1983. "A Stocker Cattle Growth Simulation Model."
    Southern Journal of Agricultural Economics 15:115-
    122.

Canas, R., J.J. Romero, and R.L. Baldwin. 1982.
    "Maintenance Energy Requirements During Lactation
    in Rats." J. Nutr. 112:1876.

Ferrell, C.L., W.N. Garrett, and N. Hinman. 1976a.
    "Growth, Development and Composition of the Udder
    and Gravid Uterus of Beef Heifers During Preg-
    nancy." J. Anim. Sci. 432:1477.

Ferrell, C.L., W.N. Garrett, N. Hinman, and G. Grich-
    ting. 1976b. "Energy Utilization by Pregnant and
    Non-Pregnant Heifers." J. Anim. Sci. 42:937.

60

Ferrell, C.L., J.A. Nienaber, and L.J. Koong. 1983.
"Effect of Previous Nutrition on Maintenance
Requirements and Efficiency of Feed Utilization of
Growing Lambs." J. Anim. Sci. 57 (Suppl. 1):431
(Abstract).

Forbes, J.M. 1977a. "Interrelationships Between
Physical and Metabolic Control of Voluntary Food
Intake in Fattening, Pregnant and Lactating Mature
Sheep: A Model." Anim. Prod. 24:91.

Forbes, J.M. 1977b. "Development of a Model of Voluntary Food Intake and Energy Balance in Lactating
Cows." Anim. Prod. 24:203.

Fox, D.G., and J.R. Black. 1977. "A System for Pre-
dicting Performance of Growing and Finishing
Cattle. 1. Development of a Model to Describe
Energy and Protein Requirements and Feed Values."
Feedstuffs 48:21.

_____. "A System for Predicting Body
Composition and Performance of Growing Cattle." J.
Animal Science. 58:725.

Fox, D.G., D.K. Phillips, R.F. Weiser, and D.J. Brown.
1981. "BEEFGAIN: A Microcomputer Program for
Predicting Performance and Profitability for Growing Cattle." Unpublished memo, Department of
Animal Science, Cornell University, Ithaca, NY.

Gill, M., J.H.M. Thornley, J.L. Black, J.D. Oldham, and
D.E.Beever. 1985. "Simulation of the Metabolism
of Absorbed Energy-yielding Nutrients in Young
Sheep." Br. J. Nutr. 52:621.

Golding, E.J., J.E. Moore, D.E. Franke, and D.C.
Ruelke. 1976. "Formulation of Hay-Grain Diets for
Ruminants. II. Depression in Voluntary Intake of
Different Quality Forage by Limited Grain in
Sheep." J. Anim. Sci. 42:717.

Holmes, W., J.G.W. Jones, and R.M. Drake-Brockman.
1961. "The Feed Intake of Grazing Cattle. II.
The Influence of Size of Animal on Feed Intake."
Anim. Prod. 3:251.

Long, C.R., T.C. Cartwright, and H.A. Fitzhugh, Jr.
1975. "Systems Analysis of Sources of Genetic and
Environmental Variation in Efficiency of Beef Production: Cow Size and Herd Management." J. Anim.
Sci. 40:409.

Meissner, H.H., and C.Z. Roux. 1984. "Growth and Feed Intake Patterns. 2. Application to Feed Efficiency." In Herbivore Nutrition in the Subtropics and Tropics. F.M.C. Gilchrist and R.I. Mackie (Eds.), p. 691, The Science Press, Craighall, R.S.A.

Moe, P.W., W.P.Flatt, and H.F.Tyrrell. 1972. "The Net Energy Value of Feeds for Lactation." J. Dairy Sci. 55:945.

Montgomery, M.J., and B.R. Baumgardt. 1965. "Regulation of Food Intake in Ruminants. I. Pelleted Rations Varying in Energy Concentration." J. Dairy Sci. 48:569.

Moore, J.E. 1978. "Forage Quality and Animal Performance." Proc. Forage and Grassland Conference, American Forage and Grassland Council, Raleigh, NC, pp. 27-34.

Moore, J.E., M.A. Worrell, S.M. Abrams, W.R. Ocumpaugh, and G.O. Mott. 1981. "Quality of Tropical Perennial Grass Hays." 1981 Beef Cattle Research Report. Animal Science and Agronomy Departments, University of Florida, Gainesville, pp. 40-44.

National Research Council. 1984. Nutrient Requirements of Beef Cattle, 6th revised ed. National Academy Press, Washington, D.C.

Neson, B.D., H.D. Ellzey and E.B. Morgan. 1968. "Effects of Feeding Varying Forage to Concentrate Ratios to Lactating Dairy Cows." J. Dairy Sci. 51:626 (Abstract).

Preston, R.L. 1985. "1984 Nutrient Requirements of Beef Cattle." Proc. Florida Nutr. Conf., University of Florida, Gainesville, p. 41.

Roux, C.Z. and H.H. Meissner. 1984. "Growth and Feed Intake Patterns. I. The Derived Theory." In Herbivore Nutrition in the Subtropics and Tropics. F.M.C. Gilchrist and R.I.Mackie (Eds.), p. 672. The Science Press, Craighall, R.S.A.

Song, H. and C.A. Dinkel. 1978. "Mathematical Models of Postweaning Growth, Feed Intake and Carcass Composition of Beef Cattle. I. Empirical Predictive Model of Voluntary Feed Intake from Weaning to Slaughter." J. Anim. Sci. 47:56.

Spreen, T.H., J.A. Ross, J.W. Pheasant, J.E. Moore, and
     W.E. Kunkle. 1985. "A Simulation Model for Back-
     grounding Feeder Cattle in Florida." Bulletin No.
     850, Fla. Agr. Exp. Sta., University of Florida,
     Gainesville.

Stokes, K.W., D.E. Farris, and T.C. Cartwright. 1981.
     "Economics of Alternative Beef Cattle Genotype and
     Management/Marketing Systems." So. J. Agr. Econ.
     13:2:1-10.

IV

The Kentucky Beef-Forage Model

Otto J. Loewer and Edward M. Smith

The modeling of beef-forage systems has been an on-going activity at the University of Kentucky since 1976. The progess of this effort, as exemplified by two regional research projects involving 25 states, has been the development of several beef and forage models. These models differ somewhat in complexity, purpose and biological orientation. Hence, the objectives of this paper are to:

1. Present a historical perspective of the modeling effort.
2. Provide a brief description of all the models.
3. Provide a more detailed biological description of one particular model.

HISTORICAL PERSPECTIVE

The beef-forage modeling effort began at the University of Kentucky in 1975 with a National Science Foundation grant to determine the energy usage associated with beef production systems. The original model was called BEEF (Beef Economic Evaluator for Farms) and will be discussed in some detail later in this paper. As a result of this model, Kentucky was invited to participate in the NC:114 regional project entitled "Forage Production and Utilization Systems as a Base for Beef and Dairy Production." In 1981, the regional effort was broadened to include the project S:156 -

---

Department of Agricultural Engineering, University of Arkansas, Fayetteville, Arkansas and Department of Agricultural Engineering, University of Kentucky, Lexington, Kentucky. Published with the approval of the Director, Arkansas Agricultural Experiment Station.

"Simulation of Forage-Beef Production in the Southern Region." In total, agricultural scientists from 25 states have had input into the development of the beef-forage models.

As a result of the regional input, the plant and animal portions of the BEEF model were developed as separate models. As the sub-models developed, they were incorporated, in some instances, in the earlier logic of BEEF.

The following models have been utilized in evaluating forage-beef production systems (Figure 1):

1. BEEF - This was the first model developed and is a "total" beef-forage simulation model. It is composed of four major sections: plant, animal, economic, and energy. It is a "consequence of actions" model that has no direct optimization features. The time step is one day. All other models in this report originated from BEEF (Loewer, et al., 1980, 1981).

2. GROWIT (Version 1) - This is a model of plant growth that utilizes maximum and minimum temperature, rainfall, soil fertility, harvesting, and other cultural practices in describing plant growth. It simulates grazing pressure through various harvesting options. Forage quality is reflected by ME and DP content. Summary reports may be obtained each day (Smith ahd Loewer, 1983; Smith, et al., 1985).

3. GROWIT (Version 2) - This model differs from Version 1 mainly in how quality is defined. Version 2 is more basic to plant growth physiology and incorporates cell content and cell wall logic with plant physiological age to define potential digestibility.

4. INTERFACE - This model incorporates the feed quality factors from GROWIT (Version 2) with grazing patterns of animals so as to define dry matter intake. Considerations include eating rate, bite size, rumination time, rumen capacity, and grazing periods.

5. BABYBEEF - This model incorporates National Research Council growth equations with environmental factors and a broad measure of physiological growth. The feed quality inputs to BABYBEEF are specified in terms of metabolizable energy (ME), digestible protein (DP), and moisture content. The time step is one day (Loewer, et al., 1983a).

6. BEEF-S156 - This model is based on the concept of physiological growth as expressed through changes in body composition. Intake may be altered by control mechanisms associated with thermostatic limits, chemostatic limits, physical fill, and night-time limits. The time step is a user-defined

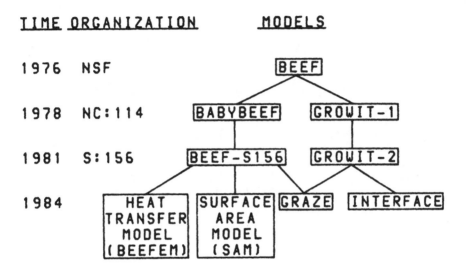

Figure 1. Forage-beef model development at the
University of Kentucky

input but is usually not greater than 15 minutes (Loewer, et al., 1983b).

7. SAM (Surface Area Model) - This model utilizes geometric shapes to describe partial surfaces of the beef animal. The purpose of this model is to enhance the ability to describe heat flows to and from the animal (Taul, 1984; Taul, et al., 1985).

8. BEFFEM - This is a heat transfer model of the interior of the beef animal. It utilizes the concepts of finite element analysis, body composition, and surface area in predicting body temperature (Turner, 1984).

9. GRAZE - This model incorporates selective grazing logic with the plant growth concepts of GROWIT (Version 2) and the animal growth concepts of BEEF-S156. The animal attempts to maximize its digestible dry matter intake rate by selecting plant material among a variable number of sub-areas within the total grazing area available to the animal (Loewer, et al., 1985).

In addition to the above references, a detailed description of several of the models may be found in Watson and Wells (1985). This reference also contains a philosophical treatment of the use of these models in addressing problems associated with forage-beef systems.

THE BEEF MODEL

The BEEF model is a total farm systems simulation model. BEEF is an interdisciplinary effort that allows users to effectively evaluate the consequences of management decisions on plant and animal production, energy consumption, and economic return.

The farm manager is influenced by many factors when determining a management strategy. These include land, money, animals, fertilizer, fuel, labor, market conditions, and other similar types of resources. The manager develops a mental image or "model" of how these resources will interact over his planning period. He then selects a set of management strategies that will best satisfy his management objectives. During his career, he may make 50 sets of management strategies. With each year of experience, he refines his model. However, he may complete his career without ever having an accurate picture of how the various system components function together. Agricultural researchers and extension specialists are faced with a similar situation. They must make decisions based on the best information available, even though their "model" may be incomplete. With this situation in mind, the computer simulation model BEEF (Beef Economic Evaluator for Farms) was developed.

The initial Kentucky BEEF model was the result of an interdisciplinary effort by a team of researchers from the Departments of Agricultural Engineering, Animal Science, Agricultural Economics and Agronomy. The project was funded in part by the National Science Foundation (Walker, et al., 1977a,b,c). The objective of the research team was to provide a management, planning and teaching tool for researchers, extension specialists and farmers, capable of determining the consequences of various management and research strategies on production.

MODELING TECHNIQUE

The BEEF model is a combination continuous-discrete simulation that utilizes FORTRAN IV with the GASP IV simulation language (Pritsker, 1974). The BEEF model contains approximately 8,000 source statements with a total length of 14,000 cards. BEEF requires more than 900 kilobytes of computer memory storage and costs approximately one-half cent per field per simulated day, depending on the number of output options. Presently, BEEF is submitted in batch form.

The systems dynamics approach is used for the continuous portions of the model such as plant and animal growth (Forrester, 1968; Walker, et al., 1977b). This approach defines daily rates of growth and utilization of a system component. The growth rate is added to the existing level of the component, while the utilization rate is subtracted. For example, the daily growth rate of a particular forage is influenced by soil fertility, soil pH, cultural practices, etc., while the utilization rate is a function of cattle size, number, and similar factors. The net effect of these rates, when added to the existing dry matter level on a particualr day, gives the dry matter level on the following day.

Discrete activities include scheduling of cultural practices, such as planting or harvesting, and the purchase or sale of resources. A discrete activity occurs instantaneously. That is, at one point in time a purchase is made or the planting of crops begins. However, the activity may continue over several days after initiation of the activity as is the case in planting a crop.

The BEEF model utilizes a "field" as the area where management activities are directed. A field is defined as a homogeneous land area where any specified management activity occurs over the entire area. For example, if the planting of a crop is scheduled, it must occur over the entire field. There is no limit as to field size or the number of fields (except in terms of array size).

Because BEEF can be used for planning future activities (e.g., a planning model), it utilizes average weather data in determining crop and animal performance, and has no stochastic processes. However, a period of stress conditions, such as drought, may be imposed by the model user to evaluate the performance of the system under adverse conditions.

BEEF is a "consequences of action" type model; that is, the user may specify almost any set of management decisions and BEEF will determine the consequences of these actions with regard to physical and economic happenings. No judgement is made concerning the desirability or logic of the management decisions; rather the philosophy of BEEF is that "desirability" will be reflected in physical performance and economic return.

The model user specifications take two forms: resources and management. The resource specifications describe the "capital" available for management at the beginning of the simulation. The manangement specifications indicate how the existing resources will be used and modified through future production, purchases, and sales. Each management specification includes the year, month, and day on which the action will occur, the work rate, field location, and other parameters that further describe the management decision. BEEF, through the GASP IV simulation language, carries out the management decisions in sequential order utilizing the resources available at that point in simulated time.

BEEF is divided into four major subsystems: crop growth, animal growth and reproduction, economic activities, and energy. These subsystems are described in great detail by Walker, et al. (1977a,b,c), including the references cited and the numeric relationships used in evaluating alternative management strategies. These relationships were entered into BEEF as "analyst input" data meaning that numeric values of the various input coefficients can be updated as better and more complete information is obtained from other research projects. The analyst input sector also allows the model user to determine the sensitivity of the model to an individual input.

CROP GROWTH

The model, as it is presently constructed, can simulate the growth of 28 different crops and crop mixes. The number and kinds of crops can easily be changed to adapt the model to any geographical region. A set of equations is used to describe the growth rate during the normal growing season for each crop planted in each field in the best soil in the given geographical region. The set is supplemented with recommended production practices. During the simulation, these growth rates are modified according to each of the various operations previously specified by the model user.

The operations that result in modification of the normal growth rates of crops are as follows:

1. Tillage operations: The simulator monitors the management input specifications for each field and records if the field has been tilled and the timing and method of tillage. This information is used to determine the appropriate growth modifier at the time when a crop is planted.

2. Planting operations: The analyst input sector stores the normal length of growing season, the time delay for emergence, and the planting date, planting method, and row spacing growth rate modifiers for each crop. When a planting operation is scheduled, the simulator selects the time when the crop will start growing and appropriate growth rate modifiers for date of planting, method of planting, and row spacing. This analyst input information would vary with geographical area.

3. Fertilizer application operations: Growth rate is also affected by the available nitrogen, phosphorus, and potassium. The use rate of each element by each crop per unit of dry matter produced, and the quantity of each element that is adequate for normal growth are stored in the analyst input data. Equations are used to calculate the loss of nitrogen due to leaching as a function of the time of the year. The simulator maintains an accounting of the quantity of each element in the soil according to the amount applied, amount used by the growing crop, and amount lost. As long as the quantity in the soil is adequate, the growth rate modifier maintains a value of 1.0, but when the quantity is below the adequate level the simulator selects an appropriate growth rate modifier.

4. Lime application operations: Growth rate modifiers for each crop for different values of soil pH are stored in the analyst input data as are the changes in soil pH per ton of lime added to different types of soil. The simulator maintains an accounting of the soil pH in each field according to the amount of lime applied and determines the appropriate growth rate modifier as a crop is growing.

5. Chemical application operations: Growth rate modifiers for weed and insects vary according to the time during the growing season when these pests affect the growth rate of each crop. The simulator records the kind of chemical specified by the model user and when it is applied on each field. If the time arrives when a certain pest will begin to

affect the growth rate of the growing crop and a chemical has not been applied that will control the pest, the simulator will select the appropriate growth rate modifier and use it during the time interval when the pest affects the crop growth.

6. Row crop cultivation operations: Weeds can be controlled by cultivation and/or herbicides. Before the simulator selects a growth rate modifier for the effect of weeds, it will check whether row crop cultivation has been performed on the field.

7. Pasture maintenance operations: Pastures may be clipped during the year to maintain the grass in the vegetative growth stage, control weeds, and reduce competition when legumes are interseeded into the pastures. The simulator adjusts the growth rate to the vegetative stage and adjusts accordingly the nutrient content of the growing crop for beef animals each time a pasture clipping operation is scheduled on a field.

8. Harvesting operations: Harvesting operations on grasses and legumes cause the growth rate and nutrient content of the growing crop for beef animals to revert to the vegetative growth stage. On the other hand, harvesting operations on grain crops cause the growth rate to become zero, and the crop dissappears from the field, leaving crop residue, which has a negative growth rate, i.e., a loss rate. The simulator adjusts the growth rate and nutrient content of the growing crop and crop residue each time a harvesting operation is scheduled to harvest a crop. The simulator also maintains an accounting of the crop that is harvested as it is placed in storage to be fed or sold. Losses of dry matter and nutrients during harvesting and storage are also accounted for by the simulator.

9. Animal moving operations: Beef animals may be moved by the model user from field to field during the year to utilize growing crops by grazing. Grazing beef animals are constantly harvesting the growing crop and consequently have an effect on the crop growth rate and nutrient content. When the quantity of growing dry matter reaches a certain minimum level, the growth rate of the crop is reduced because of the reduction in leaf area to intercept solar energy. The simulator determines the rate of consumption of growing matter by the grazing animals and maintains an accountig of the quantity of growing dry matter for each crop on each field. When the grazing animals reduce the quantity of growing dry matter below a certain minimum level, the simulator reduces the growth rate as the quantity of dry matter is reduced. The simulator maintains an accounting of the nutrient content for beef animals of the growing crop. When

the quantity of growing dry matter is reduced to a fixed minimum level by grazing animals, the nutrient content reverts to the vegetative growth stage of the growing crop. When the end of the growing season for each crop is reached, the simulator imposes a natural loss rate, i.e., a negative growth rate, which diminishes the dry matter left on each field from a previous growing season. The simulator also maintains an accounting of the number, age, sex, reproductive status, and weight of the animals that are grazing the growing crop on each field.

ANIMAL GROWTH AND REPRODUCTION

Beef animals have been identified by 12 different categories for the purpose of simulating their growth and reproduction. These categories serve to group the animals according to age, sex, and reproductive status. The simulator maintains, for each field, an accounting of each category for (1) the number of animals, (2) their weighted average age, and (3) the average weight per animal. The number of animals in each category on each field can change because of age, breeding, animal movement to other fields, birth, death and castration. For example, calves reach 12 months of age and become yearlings, non-pregnant cows are bred and become pregnant cows, animals are moved onto or away from a field, calves are born, death occurs, and castration of male animals is scheduled.

The simulator changes animals from one category to another based upon its accounting of animal age and the instructions given in the input specifications concerning breeding, animal movement to or from a field, and castration. The number of animals in each category are adjusted each time a change is made.

The simulator maintains an accounting of the average age of the animals in each category on each field by chronologically updating the age each day. The average age of the animals in each category on each field can change when animals are moved into different categories by birth of calves, breeding, movement of animals to or from a field, castration of male animals, and when animals reach an age that transfers them to another category. When animals are moved into each category on each field the simulator computes a new weighted average age for each category on each field and continues the chronological updating of age.

Breeding on a given field is initiated when yearling and/or mature bulls are scheduled by the input information to be moved onto that field with heifers and/or non-pregnant cows, and when artificial insemination is scheduled. Each category of bulls and method of artificial insemination has a characteristic breeding rate, i.e., number of females covered per day. The simulator computes a breeding rate based upon the number of bulls and cows of each category on each field. The simulator maintains an accounting of the number of heifers and non-pregnant cows that are available for breeding during the heat cycle, taking into account a time delay after calving before lactating cows are available for breeding and that yearling heifers have to reach a certain age and weight before they are available for breeding. The simulator uses the bull breeding rate, the number of females in each category available for breeding on a given day, and body composition of female animals in each available category to compute a conception rate for each female category that is available on each field for breeding. Females that conceive are moved into either the pregnant, nonlactating category and/or the pregnant, lactating category. When the females are moved into the pregnant categories, the simulator creates an unborn-calf category and maintains an accounting of the weighted average age of this category. When the age of the unborn-calf category is equal to the gestation period, the calves are born and moved into one of the calf categories.

Animal growth is represented by live weight per animal, and growth rate is represented by the rate of change in live weight as a function of time, i.e., gain or loss of weight per animal per day. The model uses input specifications describing feeding and grazing schedules and castration and health care options. Equations and analyst input data describe dry matter demand, dry matter intake, nutrient content of consumed dry matter, and the utilization of consumed dry matter for body maintenance, lactation, and gain (NRCCAN, 1969). These factors are used to simulate the change in live weight of each category of animals on each field as a function of time. The live weight per animal for each category on each field is updated each day of the simulation.

Dry matter demand is based upon the average age of the animals in each category and the potential weight per animal for animals of this age. The potential weight-age relationship for each category of animals including breed differences are stored in the program as analyst inputs. The simulator maintains an accounting of the average age of the animals in each category on each field and selects the weight per animal from the potential weight-age relationship for each category. This weight is used to compute the gross dry matter demand for each category of animals on each field each day.

Dry matter intake by each category of animals on each field is based upon the dry matter demand, the quantity of dry matter available, the metabolizable energy, and digestible protein content of the available dry matter, and the mean daily temperature and relative humidity. Selective grazing logic allows the animal to consume the highest quality forage available on a given day. Creep feeding of calves and supplemental feeding of all animals is allowed.

The dry matter that is consumed by each animal category on each field is utilized by the animals for maintenance of body functions and weight, lactation, and gain in body weight. The simulator determines each day the quantity of dry matter needed by each category of animals for each of these physiological processes and compares the need with the amount consumed. The dry matter needs are based upon metabolizable energy, digestible protein, and the metabolic body weight of the animals (Lofgreen and Garrett, 1968). If the needs for body maintenance and lactation are satisfied, then excess dry matter is utilized to provide gain in body weight. Conversely, failure to satisfy the needs for body maintenance and lactation results in loss of body weight. Analyst input data may be used to increase the dry matter needs for maintenance of body functions and weight for each category over defined time periods due to the incidence of diseases and parasites.

ECONOMIC ACTIVITIES

The BEEF model allows the user to make economic decisions in much the same way as would be done under actual farm conditions. The model user may alter the cash flow of the farm by scheduling of notes (payable or receivable) and miscellaneous income and expenses. The user may purchase or sell resources, and an income and net worth statement are prepared. Expected prices for resources that are either purchased or sold are set by the model user.

Money Accounts

BEEF contains three types of money accounts: checking, savings, and loans. Each account contains a quantity of money specified by the user. The checking account has no interest rate associated with it. The interest rates for the savings and loan accounts are specified by the user.

Only one savings account is allowed, but the user may borrow from up to seven separate loan accounts. The user may automatically control the flow of money among accounts by specifying the high and low limits of checking and savings accounts. For example, one option is to set limits on the accounts so that if the checking account is overdrawn, money is automatically trans-

ferred from savings to checking. When the savings
account is overdrawn, money flows to it from the loan
accounts.

When a purchase is made, the user must designate
the account that will be used for payment. If a loan
account is used, interest will accumulate on the unpaid
balances. Likewise, the account that will receive
money from a sale must be specified. If a loan account
receives money, the interest is paid first followed by
the principal. Any additional money from the sale
flows into the checking account.

Miscellaneous Income and Expenses

The user may specify a schedule for miscellaneous
income and expenses over the simulation period. For
example, if the model user wishes to consider income
from a part-time job as part of the cash flow in the
analysis, the quantity of money involved would be specified as would the time when the money would be added
to the user specified account.

Notes Payable and Receivable

The user may specify notes, either payable or
receivable. A schedule of payments is established by
the user including the money account to which the note
will be paid or received. Interest paid or received is
computed internally as is the present value of the
note. The notes payable and receivable section allows
the user to account for debts incurred before the simulation period begins.

Purchases

All the resources required for production may be
purchased at times and for amounts specified by the
user. In addition, the user may choose to defer payments using a uniform payment series with or without a
down payment. All purchases must be charged to one of
the money accounts. The delivery day of the purchased
item is also specified. When this date is reached by
the simulation, the inventory of this item is adjusted
to reflect the total quantity of the item on hand and
its value.

Sales

Resources may be sold at user specified prices at
any point during the simulation period. The quantity

sold is expressed as a percentage of the supply that is in inventory at the time of the sale. For example, the user may specify that 50 percent of the yearling heifers on Field No. 3 will be sold on May 20 of the simulation period for $50.00 per hundredweight, and the money from the sale will be placed in Account No. 3 (the loan account). On May 20 of the simulation period, 50 percent of the yearling heifers are removed from the Field No. 3 inventory. The total weight of these animals (computed internally by BEEF) is multiplied by the price and the sum of money is transferred to Account No. 3, first to pay the interest due, then to repay the principal. Excess funds, if any, go into the checking account.

Income Statement

At the end of each simulated year two different forms of an income statement are prepared. One is based on actual expenses while the other is based on required expenses. The statements are developed by internally determining the total expenses and receipts over the simulated year using the cash basis of accounting. Depreciation for machinery is calculated based on user supplied depreciation method data, and is considered a deductible expense. The primary difference between this income statement and the standard Internal Revenue Service technique is that no personnel deductions are allowed for items such as the number of dependents, contributions to charity, etc.

Net Worth

Taxable income is one indicator of profitability with net worth being another. An inventory of all resource items and their value per unit are maintained internally. The net worth statement reflects the values of the expendable items in inventory, the market value of machinery, the market value of land, the present value of notes and loans, and cash on hand. The change in net worth from the beginning of the simulation is also maintained.

ENERGY

The BEEF model maintains an accounting of the energy use incurred on a daily basis for owning and maintaining buildings, fences, roads, and the energy use for each production operation as it occurs during the simulation. The energy use for production operations is identified according to the amount (Kcal) used for fuel; labor; manufacture, transport, and repair of machines; seed; fertilizer; lime; chemicals; protein supplement; and mineral supplement. Analyst input data provide energy values for different methods of performing production operations and materials used in these operations; input specifications describe when and how each operation will be performed. A complete description of the techniques used in BEEF is given by Bridges and Smith (1979).

OUTPUT

The BEEF model has several forms of output that may be selected by the model user. These include tables, event monitoring, and plots with optional tables.

Tables

BEEF has a standard set of output tables that reflect the initial conditions of the system. If other output information is desired, it must be specified by the user. Additional tables in any combination may be printed with any frequency. The types of table information include land, machinery, money accounts, notes payable, notes receivable, cattle, fertilizer, lime, chemicals, fuel, seed, protein, minerals, silage, haylage, hay, grain, cubes and wafers, growing crops, present worth, and energy factors. There is a summary table for each year and for the total simulation period that provides statistical ratios and performance data for each category of animal.

Event Monitoring

The BEEF model allows the user to monitor the management decisions (events) that occur over the simulation period. Field operations and herd management operations may be traced over any time period for any number of fields or categories of beef animals. Financial transactions, such as purchases or sales, may also be monitored.

Plots with Associated Tables

FORTRAN plots with or without associated tables may be selected for energy utilization, cash flow, pasture performance, and animal performance. The plot for energy utilization shows the components of the energy used in addition to the total quantity used over time. This allows the user to monitor any category of energy consumption, such as fuel, over the simulation period. The cash flow plot monitors the checkings, savings, and loan accounts, and the total cash available. Interest on the loan account is also tabulated. Pasture performance may be monitored up to four fields. The dry matter level and rate of growth are presented in addition to other factors. However, no animal performance information is presented in these plots. Up to four animal performance plots may be obtained. Each category of animals is monitored on a user specified field rather than added together for the cumulative effect over the total farm.

Sample Output

The following figures are representative of the types of output that may be obtained from BEEF. The model begins by listing all the activities scheduled at the beginning of the simulation (Figure 2). A display of all the cattle breed characteristics is provided (Figure 3) as is an accounting of all production resources. For example, the initial land resource base is shown in Figure 4. Production resources are updated each day. The cattle and field dry matter status for a particular day are shown in Figures 5 and 6. At the end of any simulated year, several forms of summary output are given. Income statements for both required and scheduled transactions are given (Figure 7) as is a financial statement (Figure 8) and financial ratio analysis (Figure 9). Summary statistics are also provided by BEEF for a given period of simulated time (Figures 10-12). Graphical displays (not shown) are also available showing daily changes in plant and animal performance.

USER INPUTS

An input booklet was prepared for the BEEF user to assist him in using the model. The book contains a complete set of input forms with detailed instructions for data entry. Each input sheet utilizes a standard FORTRAN input form so that card images may be entered directly. The complete book will require approximately 8 hours to complete depending on farm size and the complexity of management strategies. However, changes in resources or basic management strategies will require only minutes to input once the basic farm data is available to be used by the computer.

****
****

LISTING OF ALL MANAGEMENT DECISIONS, PURCHASES AND SALES

****

ENTRY	TIME	EVENT	EVENT CODE DESCRIPTION	ATTRIB. 3	ATTRIB. 4	ATTRIB. 5	ATTRIB. 6	ATTRIB. 7
	DAY MNTH YEAR	CODE		ATTRIB. 8	ATTRIB. 9	ATTRIB. 10	ATTRIB. 11	ATTRIB. 12
				ATTRIB. 13	ATTRIB. 14	ATTRIB. 15	ATTRIB. 16	ATTRIB. 17
				ATTRIB. 18	ATTRIB. 19	ATTRIB. 20	ATTRIB. 21	ATTRIB. 22
				ATTRIB. 23	ATTRIB. 24	ATTRIB. 25		
1	0.0	222.1	FEEDING HAY	0.1000E 01	0.500E 00	0.1200E 02	0.211BE 02	0.20000E 01
	1 JAN., 1980			0.5200E 02	0.0	0.0	0.0	0.0
				0.0	0.0	0.0	0.0	0.0
				0.0	0.0	0.0	0.0	0.0
				0.0	0.0	0.0		

****

DEFINITION OF ATTRIBUTES
MANAGEMENT EVENTS

FOR FIELD OPERATIONS AND HARVESTING---(3) FIELD ID; (4) WORK RATE; (5) CUSTOM OPERATION ID.; (6) CUSTOM OPERATION COST (/$/AC).
FOR FEEDING OPERATIONS-----------------(3) FIELD ID; (4) DIST. FROM STORAGE TO FIELD ; (6) AMOUNT TO BE FED EACH DAY (LB/AN)
                                       (7) FEEDING SCHEDULE ID.; (8) REPETITIONS OF FEEDING SCHEDULE.
FOR CATTLE OPERATIONS------------------(3) ANIMAL CAT. ID.; (4) PERCENT OF CAT. TREATED; (5) CUSTOM OPER. ID.; (6) CUSTOM
                                       OPERATIONS COST ($/AN).

PURCHASE EVENTS--INVENTORY ADJUST

FOR ANIMALS------(3) CAT. ID.; (5) NUMBER PURCHASED; (6) AVERAGE AGE (MONTHS); (7) WEIGHT (LB); (8) FIELD ID.; (9) COST ($/100 LB);
                 (10) DAYS SINCE BRED (CAT. 7 & 9); (11) DAYS SINCE CALVES DROPPED (CAT. 8 & 9); (12) BREED ID; (13) HOW ACTIVE:
                 (14) HAIR LENGTH (CM); (15) DAIRY (1=YES).
FOR GRAIN--------(3) CROP ID.; (4) AMOUTN (BU); (5) COST/BU.
FOR FERTILIZER---(3) FERTILIZER ID.; (4) AMOUNT (TONS); (5) COST/TON.
FOR LIME---------(3) ID.; (4) AMOUNT (TONS); (5) COST/TON.
FOR SEED---------(3) ID.; (4) AMOUNT (B); (5) COST/LB.
FOR LAND---------(3) ID.; (4) AMOUNT (ACRES); (5) COST/ACRE.
FOR THE REST-----(3) ID.; (4) AMOUNT; (5) COST/UNIT.

PURCHASE EVENTS--MONEY ACCOUNT ADJUST

FOR ALL----------(3) PAID FROM MONEY ACCOUNT ID.; (4) ACTUAL $ SPENT TODAY; (5) COST/UNIT.

SALES EVENTS

FOR ANIMALS------(3) CAT. ID.; (4) % OF CAT. SOLD; (5) PRICE ($/100 LB); (7) FIELD ID.
FOR OTHER--------SAME AS PRUCHASES -INV. ADJ.
****

Figure 2. Example output listing by BEEF of initially scheduled management decisions and purchases and sales.

CATTLE BREED CHARACTERISTICS

****

BREED ID*	ATTRIBUTE NUMBER	ATTRIBUTE NAME	SEX*	VALUE OF ATTRIBUTE	SEX*	VALUE OF ATTRIBUTE	SEX*	VALUE OF ATTRIBUTE
1	1	WTMI	1	0.7500	2	0.7500	3	0.7500
1	2	PASSDA	1	3.0000	2	3.0000	3	3.0000
1	3	EFATKG	1	7500.0000	2	7500.0000	3	7500.0000
1	4	EFFFAT	!	35.0000	2	35.0000	3	35.0000
1	5	ETISKG	1	2500.0000	2	2500.0000	3	2500.0000
1	6	PCCEQT	1	0.0	2	0.0	3	0.0
1	7	HAIRMN	1	0.5000	2	0.5000	3	0.5000
1	8	HAIRMX	1	3.0000	2	3.0000	3	3.0000
1	9	HAIRGR	1	0.1000	2	0.1000	3	0.1000
1	10	HAIRDR	1	0.2000	2	0.2000	3	0.2000
1	11	SURFXX	1	0.0900	2	0.0900	3	0.0900
1	12	RESMIN	1	0.0084	2	0.0084	3	0.0084
1	13	RESMAX	1	0.1140	2	0.1140	3	0.1140
1	14	PCSKBO	1	50.0000	2	50.0000	3	50.0000
1	15	PCMWDR	1	0.5000	2	0.5000	3	0.5000
1	16	RECTLT	1	38.0000	2	38.0000	3	38.0000
1	17	RMIN	1	0.0	2	0.0	3	0.0
1	18	PCADMN	1	100.0000	2	100.0000	3	100.0000
1	19	PCSATN	1	100.0000	2	100.0000	3	100.0000
1	20	PCADMX	1	100.0000	2	100.0000	3	100.0000
1	21	PCSATX	1	100.0000	2	100.0000	3	100.0000
1	22	DAYCB	1	281.0000	2	281.0000	3	281.0000
1	23	DAYBMR	1	277.0000	2	277.0000	3	277.0000
1	24	DMRGM	1	1177.0000	2	1177.0000	3	1230.0000
1	25	ADLTBB	1	0.0210	2	0.0130	3	0.0230
1	26	ADLTRR	1	0.7600	2	1.1500	3	0.5900
1	27	GRMAT	1	0.500E-08	2	-.250E-08	3	0.100E-09
1	28	XMXMLK	1	0.0	2	0.0	3	0.8400
1	29	XMMEKG	1	0.0	2	0.0	3	6200.0000
1	30	EFFMLK	1	0.0	2	0.0	3	65.0000
1	31	STRMLK	1	0.0	2	0.0	3	3.0000

Figure 3: Example output listing by Beef of cattle breed characteristics (1 of 4)

#	Variable		1		2		3
32	XMINMK	1	0.0	2	0.0	3	0.1000
33	OILKDA	1	0.0	2	0.0	3	7.0000
34	PCDRYM	1	0.0	2	0.0	3	12.0000
35	POSTPM	1	0.0	2	0.0	3	45.0000
36	PCBRED	1	0.0	2	0.0	3	65.0000
37	RABFET	1	0.0	2	0.0	3	1.0000
38	XMLKDP	1	0.0	2	0.0	3	27.8000
39	XMLKC	1	0.0	2	0.0	3	1.2600
40	XMLKT	1	0.0	2	0.0	3	1.0300
41	HAIRLB	1	3.0000	2	3.0000	3	3.0000
42	BULPRY	1	0.0	2	0.2500	3	0.0
43	BULPRM	1	0.0	2	2.0000	3	0.0
44	HMNWTB	1	0.0	2	0.0	3	209.9564
45	PCMKMN	1	0.0	2	0.0	3	100.0000
46	CNMAX	1	0.0	2	0.0	3	0.6300
47	EKLWRA	1	80.0000	2	80.0000	3	72.0000
48		1	0.0	2	0.0	3	0.0
49		1	0.0	2	0.0	3	0.0
50		1	0.0	2	0.0	3	0.0
1	WTMI	1	0.7500	2	0.7500	3	0.7500
2	PASSDA	1	3.0000	2	3.0000	3	3.0000
3	EFATKG	1	7500.0000	2	7500.0000	3	7500.0000
4	EFFAT	1	35.0000	2	35.0000	3	35.0000
5	ETISKG	1	2500.0000	2	2500.0000	3	2500.0000
6	PCCEQT	1	0.0	2	0.0	3	0.0
7	HAIRMN	1	0.500	2	0.500	3	0.5000
8	HAIRMX	1	3.0000	2	3.0000	3	3.0000
9	HAIRGR	1	0.1000	2	0.1000	3	0.1000
10	HAIRDR	1	0.2000	2	0.2000	3	0.2000
11	SURFXX	1	0.0900	2	0.0900	3	0.0900
12	RESMIN	1	0.0084	2	0.0084	3	0.0084
13	RESMAX	1	0.1140	2	0.1140	3	0.1140
14	PCSKBO	1	50.0000	2	50.0000	3	50.0000
15	PCMWDR	1	0.5000	2	0.5000	3	0.5000
16	RECTLT	1	38.0000	2	38.0000	3	38.0000
17	RMIN	1	0.0	2	0.0	3	0.0
18	PCADMN	1	100.0000	2	100.000	3	100.0000
19	PCSATN	1	100.0000	2	100.0000	3	100.0000

Figure 3 (2 of 4)

#	Variable	1	2	3
20	PCADMX	100.0000	100.0000	100.0000
21	PCSATX	100.0000	100.0000	100.0000
22	DAYCB	281.0000	281.0000	281.0000
23	DAYBMR	277.0000	277.0000	277.0000
24	DMRGM	1177.0000	1177.0000	1230.0000
25	ADLTBB	0.0210	0.0130	0.0230
26	ADLTRR	0.7600	1.1500	0.5900
27	GRMAT	0.500E-08	0.250E-08	0.100E-09
28	XMXMLK	0.0	0.0	0.8400
29	XMMEKG	0.0	0.0	6200.0000
30	EFFMLK	0.0	0.0	65.0000
31	STRMLK	0.0-	0.0	3.0000
32	XMINMK	0.0	0.0	0.1000
33	OILKDA	0.0	0.0	7.0000
34	PCDRYM	0.0	0.0	12.0000
35	POSTPM	0.0	0.0	45.0000
36	PCBRED	0.0	0.0	65.0000
37	RABFET	0.0	0.0	1.0000
38	XMLKDP	0.0	0.0	27.8000
39	XMLKC	0.0	0.0	1.2600
40	XMLKP	0.0	0.0	1.0300
41	HAIRLB	3.0000	3.0000	3.0000
42	BULPRY	0.0	0.2500	0.0
43	BULPRM	0.0	2.0000	0.0
44	HMNWTB	0.0	0.0	209.9564
45	PCMKMN	0.0	0.0	100.0000
46	CNMAX	0.0	0.0	0.6300
47	EWLWRA	80.0000	80.0000	72.0000
48		0.0	0.0	0.0
49		0.0	0.0	0.0
50		0.0	0.0	0.0

BREED ID: 1) ENGLISH   SEX: 1) STEER
          2)   2) BULL
          3)   3) COW

Figure 3 (3 of 4)

****
BREED CHARACTERISTICS DEFINITIONS

1	WTMI	ADJUSTED METABOLIC WEIGHT COEFFICIENT
2	PASSDA	DAYS FOR PASSAGE OF RATION
3	EFATKG	ENERGY CONTENT OF BODY FAT, KCAL/KG
4	EFFFAT	EFFICIENCY OF ADDING FAT ALONG, PERCENT
5	ETISKG	ENERGY IN TISSUE, KCAL/KG
6	PCCEQT	% ALTERATION OF NET ENERGY EQUATIONS
7	HAIRMN	MINIMUM POSSIBLE HAIR LENGTH, CM
8	HAIRMX	MAXIMUM PSSSIBLE HAIR LENGTH, CM
9	HAIRCR	GROWTH RATE OF HAIR, CM/DAY
10	HAIRDR	MAXIMUM RATE OF HAIR REMOVAL, CM/DAY
11	SURFXX	SURFACE AREA MULTIPLIER (SURFAC*AREA**0.67)
12	RESMIN	MINIMUM RESPIRATION RATE, M**3/MIN-M**2 OF SURFACE
13	RESMAX	MAXIMUM RESPIRATION RATE, M**3/MIN-M**2 OF SURFACE
14	PCSKBO	PERCENT OF POTENTIAL WEIGHT BELOW WHICH DEATH OCCURS
15	PCMWDR	MAXIMUM REMOVAL RATE OF BODY STORES, PERCENT/DAY
16	RECTLT	ANIMAL BODY TEMPERATURE, C
17	RMIN	MINIMUM INTERNAL HEAT RESISTANCE, C-M**2/W
18	PCADMN	BODY-AMBIENT TEMPERATURE DIFFERENCE OBTAINED AT RESMIN, %
19	PCSATN	BODY-AMBIENT MOISTURE SATURATION DIFFERENCE OBTAINED AT RESMIN, %
20	PCADMX	BODY-AMBIENT TEMPERATURE DIFFERENCE OBTAINED AT RESMAX, %
21	PCSATX	BODY-AMBIENT MOISTURE SATURATION DIFFERENCE OBTAINED AT RESMAX, %
22	DAYCB	DAYS FROM CONCEPTION TO BIRTH
23	DAYBMR	DAYS FROM BIRTH TO MAX. RATE OF GAIN
24	DMRGM	DAYS-MAX. RATE OF GAIN TO MATURITY
25	ADLTBB	RATE OF GAIN AT CONCEPTION, KG/DAY
26	ADLTRR	MAXIMUM POTENTIAL RATE OF GAIN, KG/DAY
27	GRMAT	RATE OF GAIN AT MATURITY, KG/DAY (MUST BE NON-ZERO)
28	XMXMLK	MAXIMUM MILK PRODUCTION POTENTIAL (DRY WEIGHT), KG/DAY
29	XMMEKG	ENERGY CONTENT OF DRY MILK, ME/KG
30	EFFMLK	EFFICIENCY OF MILK PRODUCTION, %
31	STRMLK	DAYS OF STRESS BEFORE LCATATION IS TERMINATED
32	XMINMK	MINIMUM MILK PRODUCTION, KG/DAY, ENDING LACTATION
33	OILKDA	DAYS FOR AVERAGE PRODUCTION TRENDS (1)
34	PCDRYM	PERCENT DRY MATERIAL IN LIQUID MILK
35	POSTPM	POST-PARTUM TIME, DAYS
36	PCBRED	WEIGHT BELOW WHICH THERE IS NO CONCEPTION, % OF POTENTIAL
37	RABFET	RATIO OF PLACENTA WEIGHT TO CALES WEIGHT AT BIRTH
38	XMLKDP	DIGESTABLE PROTEIN OF DRY MILK, %
39	XMLKC	CALCIUM CONTENT OF DRY MILK, %
40	XMLKP	PHOSPHORUS CONTENT OF DRY MILK, %
41	HAIRLB	HAIR LENGTH AT BIRTH, CM
42	BULPRY	YEARLING BULL (CAT. 3) BREEDING RATE
43	BULPRM	MATURE BULL (CAT. 1) BREEDING RATE
44	HMNWTB	MINIMUM WEIGHT FOR BREEDING (NORMAL WEIGHT FOR A 1 YEAR OLD FEMALE), KG
45	PCMKMN	PERCENT OF POTENTIAL WEIGHT BELOW WHICH LACTATION STOPS
46	CNMAX	MAXIMUM CONCEPTION RATE (0= CNMAX =1)
47	EWLWRA	EMPTY BODY/LIVE WT. RATIO (%)

****

Figure 3 (4 of 4)

****
****

LAND ATTRIBUTES

FIELD ID (NO.)	FIELD AREA (ACRES)	LAND USE CLASS NO.	SOIL KIND ID NO.*	ESTIMATE OF YIELD	LAND VALUE MARKET (S/A)	TAX (S/A)	TAX RATE (S/$10.00)	MONTH TAX PAYABLE	LAND VALUE MARKET ($)	TAX ($)
1	40.00	4.0	2.0	0.0	800.00	800.00	1.93	12.0	32000.00	32000.00
2	40.00	4.0	2.0	0.0	800.00	800.00	1.93	12.0	32000.00	32000.00
3	40.00	4.0	2.0	0.0	800.00	800.00	1.93	12.0	32000.00	32000.00

* 1 = SANDY OR SANDY LOAM; 2 = SILT LOAM OR CLAY LOAM; 3 = HEAVY CLAY.

FIELD ID (NO.)	INITIAL SOIL TEST DATA PH	P(LB)	K(LB)	CROP ID*	PRESENTLY EXISTING QUANTITIES N(LB)	P(LB)	K(LB)	PH	FENCED? YES = 1, NO=0	ROW SPACE (IN.)
1	7.00	100.00	250.00	23.0	0.0	100.00	250.00	7.00	1.00	0.0
2	7.00	100.00	250.00	23.0	0.0	100.00	250.00	7.00	1.00	0.0
3	7.00	100.00	250.00	23.0	0.0	100.00	250.00	7.00	1.00	0.0

* 1. NO CROP; 2. FESCUE; 3. BLUEGRASS; 4. ORCHARD GRASS; 5. TIMOTHY; 6. ALFALFA; 7. RED CLOVER; 8. LADING WHITE CLOVER
9. LESPEDEZA; 10. FESCUE ALFALFA; 11. BLUEGRASS-ALFALFA; 12. ORCHARD-GRASS-ALFALEA; 13. FESCUE = RED CLOVER;
14. BLUEGRASS-RED CLOVER; 15. ORCHARD GRASS-RED CLOVER; 16. FESCUE-LADING WHITE CLOVER; 17. BLUEGRASS-LADING WHITE CLOVER;
18. ORCHARD GRASS-LADING WHITE CLOVER; 19. FESCUE-LESPEDEZA; 20. BLUEGRASS-LESPEDEZA; 21. ORCHARD GRASS-LESPEDEZA;
22. WHEAT; 23. RYE; 24. OATS; 25. BARLEY; 26. SOYBEANS; 27. SUDEX; 28. TOBACCO; 29. CORN.

****

Figure 4. Example output listing by BEEF of initial land resource attributes

```
*************************************************************************
*************************************************************************
```

Line 1 — animal / market data

CATTLE	FIELD ID	CATEGORY ID	NUMBER OF ANIMALS	AVERAGE AVE(MO)	AVERAGE WT.(LB)	MRKT. VALUE ($/100 LB)	NUMBER SOLD	NUMBER PURCHASED	SALES ($)	PURCHASES ($)
	1	1	0.0	0.0	0.0	55.00	0.0	4.40	0.0	3872.00
	2	1	4.40	25.35	0.0	55.00	0.0	0.0	0.0	0.0
	2	8	16.28	66.93	1033.52	35.00	0.0	0.0	0.0	0.0
	2	9	93.72	66.93	1035.98	40.00	0.0	0.0	0.0	0.0
	2	10	55.00	3.55	213.87	55.00	0.0	0.0	0.0	0.0

Line 2 — breed / weight / DM demand

BREED ID	HAIR LG. (CM)	NORMAL WT.(LB)	NORMAL WT. CHANGE(LB)	ACTUAL WT.* CHANGE(LB)*	AVAIL DM	LB/ANIMAL-DAY DM DEMAND	DMI
1.00	0.0	0.0	0.0	0.0	0.0	0.0	0.0
1.00	0.60	0.0	0.10	1.33	845.38	35.07	35.07
1.00	0.50	905.24	0.00	0.44	526.29	21.83	21.83
1.00	0.50	905.25	0.00	0.60	527.49	21.88	21.88
1.00	3.00	212.26	1.34	2.08	142.53	7.58	7.58

Line 3 — milk / feed / climate

MLK INTAKE CAT(10,11,12) (LB)	CALF MILK DEMAND LB CAT(7,9)	AVE MILK PROD LB CAT(8,9)	MLK PROD TODAY LB CAT(8,9)	FEED ME KCAL/KG	PCT. PROT OF FEED	AVE TEMP TODAY DEGREES-F	AVE REL. HUMIDITY (%)
0.0	0.0	1653.75	0.0	0.0	0.0	75.76	65.00
0.00	0.0	1471.73	0.0	2600.67	12.58	75.54	65.00
0.0	1.85	1.85	1.85	2600.67	12.58	75.54	65.00
0.0	1.85	1.85	1.85	2600.67	12.58	75.54	65.00
1.74	0.0	0.0	0.0	3428.16	16.08	75.54	65.00

Line 4 — dry matter (LB/AN-DAY)

	AVAILABLE DRY MATTER				DRY MATTER INTAKE			
	PASTURE	STORED DM	CREEP	MILK	PASTURE	STORED DM	CREEP	MILK
	0.0	0.0	0.0	0.0	0.0	0.0	0.0	0.0
	840.51	4.86	0.0	0.0	34.87	0.20	0.0	0.0
	523.26	3.03	0.0	0.0	21.71	0.13	0.0	0.0
	524.46	3.03	0.0	0.0	21.76	0.13	0.0	0.0
	139.98	0.81	0.0	1.74	5.81	0.03	0.0	1.74

```
2    12    55.00   3.55    223.80  65.00    0.0     0.0     0.0     0.0
            1.00    3.00    223.16   1.50    0.60   160.15   8.52    8.52
            1.96    0.0      0.0     0.0   3428.16   16.08   75.54   65.00
          157.28    0.91     0.0     1.96     6.52    0.04    0.0     1.96

2    13    93.72    0.88     1.23    0.0      0.0     0.0     0.0     0.0
            1.00    0.0      1.23    0.06     0.06    0.0     0.0     0.0
            0.0     0.0      0.0     0.0      0.0     0.0    48.67   65.00
            0.0     0.0      0.0     0.0      0.0     0.0     0.0     0.0
```

* 1. MATURE BULLS; 2. MATURE STEERS; 3. YEARLING BULLS; 4. YEARLING STEERS; 5. NON-LACTATING
NON-PREGNANT COWS; 6. YEARLING HEIFERS NON-PREGNANT; 7. PREGNANT NON-LACTATING COWS;
11. BULL CALVES; 12. STEER CALVES; 13. UNBORN CALVES.

**THE MILK AVAILABLE TO THE CALVES LAGS THE MILK PRODUCTION OF THE COWS(CAT. 8, CAT. 9) BY ONE DAY.

Figure 5.   Example output listing by BEEF Cattle Status on a particular simulated day

```
***************************************************************************
***************************************************************************
```

FIELD DRY MATTER SECTION

```
***************************************************************************
***************************************************************************
```

DRY MATTER CATEGORIES	FIELD ID.	CROP ID.	***FROM FIELD*** FEEDING RATE, LB/D	INTAKE DM,LBS	% INTAKE TO TOTAL	STORED DM LEVEL, LBS	***FROM CREEP*** FEEDING RATE, LB/D	INTAKE DM,LBS	% INTAKE TO TOTAL	STORED DM LEVEL, LBS	INTAKE DM (LB)	*TOTAL* INTAKE DM (LB)
HAY	1	12	0.0	0.0	0.0	106.15	0.0	0.0	0.0	0.0	0.0	0.0
PASTURE	1	23	467.56	0.0	0.0	57948.23	0.0	0.0	0.0	0.0	0.0	0.0
TOTAL	1		467.56	0.0	0.0	58054.38	0.0	0.0	0.0	0.0	0.0	0.0
CUMULATIVE	1		544284.	485912.			0.					485912.
HAY	2	12	0.0	19.97	0.61	481.23	0.0	0.0	0.0	0.0	0.0	19.97
PASTURE	2	23	467.56	3229.37	99.39	77810.63	0.0	0.0	0.0	0.0	0.0	3229.37
TOTAL	2		467.56	3249.34	100.00	78291.81	0.0	0.0	0.0	0.0	0.0	3249.34
CUMULATVE	2		554378.	479337.			0.					479337.
HAY	3	12	0.0	0.0	0.0	374.49	0.0	0.0	0.0	0.0	0.0	0.0
PASTURE	3	23	467.56	0.0	0.0	71954.69	0.0	0.0	0.0	0.0	0.0	0.0
TOTAL	3		467.56	0.0	0.0	72329.13	0.0	0.0	0.0	0.0	0.0	0.0
CUMULATIVE	3		526458.	454130.			0.					454130.
TOTAL FOR ALL FIELDS			1403.	3249.		0.						3249.
CUMULATIVE-ALL FIELDS			1625118	208675.		0.						1419378.

Figure 6. Example output listing by BEEF of field dry matter status for a particular simulated day.

INCOME STATEMENT
JAN. 1, 1982

CATEGORY OF EXPENSE/INCOME	EXPENSES AMOUNT	EXPENSES PERCENT OF TOTAL	EXPENSES PERCENT CHANGE	REQUIRED TRANSACTIONS INCOME AMOUNT	INCOME PERCENT OF TOTAL	INCOME PERCENT CHANGE	NET AMOUNT	NET PERCENT OF TOTAL	NET PERCENT CHANGE
1 TAXES	185.28	0.77	0.0	0.0	0.0	-0.0	-185.28	-6.07	0.0
2 LABOR	842.29	3.49	-0.8	0.0	0.0	-0.0	-842.29	-18.50	0.8
3 INSURANCE	0.0	0.0	-0.0	0.0	0.0	-0.0	0.0	0.0	-0.0
4 INTEREST	0.0	0.0	-100.0	3135.93	10.94	433.9	3135.93	68.88	648.3
5 CUST. OPER.	1743.15	7.23	-0.5	0.0	0.0	-0.0	-1743.15	-38.29	-0.5
6 HEALTH PRTC.	0.0	0.0	-0.0	0.0	0.0	-0.0	0.0	0.0	-0.0
7 CATTLE	3872.00	16.05	0.0	25534.71	89.06	-4.6	21662.71	475.81	-5.4
8 CRAIN	0.0	0.0	-0.0	0.0	0.0	-0.0	0.0	0.0	-0.0
9 FORAGE	13815.79	57.28	-0.2	0.0	0.0	-0.0	-13815.79	-303.46	-0.2
10 FERTILIZER	3607.20	14.96	0.0	0.0	0.0	-0.0	-3607.20	-79.23	0.0
11 LIME	0.0	0.0	-0.0	0.0	0.0	-0.0	0.0	0.0	-0.0
12 SEED	0.0	0.0	-0.0	0.0	0.0	-0.0	0.0	0.0	-0.0
13 CHEMICALS	0.0	0.0	-0.0	0.0	0.0	-0.0	0.0	0.0	-0.0
14 FUEL	52.10	0.22	-2.1	0.0	0.0	-0.0	-52.10	-1.14	-2.1
15 PROTEIN	0.0	0.0	-0.0	0.0	0.0	-0.0	0.0	0.0	-0.0
16 MINERALS	0.0	0.0	-0.0	0.0	0.0	-0.0	0.0	0.0	-0.0
17 VACCINE	0.0	0.0	-0.0	0.0	0.0	-0.0	0.0	0.0	-0.0
18 MEDICIN	0.0	0.0	-0.0	0.0	0.0	-0.0	0.0	0.0	-0.0
19 SEMEN	0.0	0.0	-0.0	0.0	0.0	-0.0	0.0	0.0	-0.0
20 REPAIRS	0.0	0.0	-0.0	0.0	0.0	-0.0	0.0	0.0	-0.0
21 UTILITIES	0.0	0.0	-0.0	0.0	0.0	-0.0	0.0	0.0	-0.0
22 LAND	0.0	0.0	-0.0	0.0	0.0	-0.0	0.0	0.0	-0.0
23 MISC	0.0	0.0	-0.0	0.0	0.0	-0.0	0.0	0.0	-0.0
TOTALS FOR EXPENSE, INCOME	24117.81	100.00	-0.9	28670.64	100.00	4.8	4552.83	100.00	50.6
NET INCOME									

*PERCENT CHANGE FROM LAST YEARS VALUE. A "-0.0" INCICATES A ZERO VALUE IN DENOMINATOR.
**REQUIRED LABOR FOR ACTUAL USER APPLIED OPERATIONS WERE 240.66 HOURS AT A COST OF
$ 842.29 WITH AN EFFECTIVE HOURLY LABOR RATE OF $_____ 3.50 PER HOUR.

Figure 7. Example output listing by BEEF of income statements for required and scheduled transactions (1 of 2).

SCHEDULED TRANSACTIONS

CATEGORY OF EXPENSE/INCOME	EXPENSES AMOUNT	PERCENT OF TOTAL	PERCENT CHANGE*	INCOME AMOUNT	PERCENT OF TOTAL	PERCENT CHANGE*	NET AMOUNT	PERCENT OF TOTAL	PERCENT CHANGE*
1 TAXES	185.28	3.69	0.0	0.0	0.0	0.0	-185.28	-0.78	0.0
2 LABOR	0.35	0.01	0.0	0.0	0.0	0.0	-0.35	-0.00	0.0
3 INSURANCE	0.0	0.0	-0.0	0.0	0.0	-0.0	0.0	0.0	-0.0
4 INTEREST	0.0	0.0	-100.0	3135.93	10.94	433.9	3135.93	13.26	648.3
5 CUST. OPER.	1743.15	34.68	-0.5	0.0	0.0	-0.0	-1743.15	-7.37	-0.5
6 HEALTH PRTC.	0.0	0.0	-0.0	0.0	0.0	-0.0	0.0	0.0	-5.2
7 CATTLE	3097.60	61.63	0.0	25534.71	89.06	-4.6	22437.11	94.89	-0.0
8 GRAIN	0.0	0.0	-0.0	0.0	0.0	-0.0	0.0	0.0	-0.0
9 HAY	0.0	0.0	-0.0	0.0	0.0	-0.0	0.0	0.0	-0.0
10 SILAGE	0.0	0.0	-0.0	0.0	0.0	-0.0	0.0	0.0	-0.0
11 HAYLAGE	0.0	0.0	-0.0	0.0	0.0	-0.0	0.0	0.0	-0.0
12 FERTILIZER	0.0	0.0	-0.0	0.0	0.0	-0.0	0.0	0.0	-0.0
13 LIME	0.0	0.0	-0.0	0.0	0.0	-0.0	0.0	0.0	-0.0
14 SEED	52.10	0.0	-0.0	0.0	0.0	-0.0	0.0	0.0	-0.0
15 CHEMICALS	0.0	0.0	-0.0	0.0	0.0	-0.0	0.0	0.0	-0.0
16 FUEL	0.0	0.0	-0.0	0.0	0.0	-0.0	0.0	0.0	-0.0
17 PROTEIN	0.0	0.0	-0.0	0.0	0.0	-0.0	0.0	0.0	-0.0
18 MINERALS	0.0	0.0	-0.0	0.0	0.0	-0.0	0.0	0.0	-0.0
19 VACCINE	0.0	0.0	-0.0	0.0	0.0	-0.0	0.0	0.0	-0.0
20 MEDICINE	0.0	0.0	-0.0	0.0	0.0	-0.0	0.0	0.0	-0.0
21 SEMEN	0.0	0.0	-0.0	0.0	0.0	-0.0	0.0	0.0	-0.0
22 REPAIRS	0.0	0.0	-0.0	0.0	0.0	-0.0	0.0	0.0	-0.0
23 UTILITIES	0.0	0.0	-0.0	0.0	0.0	-0.0	0.0	0.0	-0.0
24 LAND	0.0	0.0	-0.0	0.0	0.0	-0.0	0.0	0.0	-0.0
25 MISC.	0.0	0.0	-0.0	0.0	0.0	-0.0	0.0	0.0	-0.0
TOTALS FOR EXPENSE, INCOME	5026.38	100.00	-3.4	28670.64	100.00	4.8	23644.26	100.00	6.7
NET INCOME									

*PERCENT CHANGE FROM LAST YEARS VALUE. A "-0.0" INDICATES A ZERO VALUE IN DENOMINATOR.
**SCHEDULED LABOR REQUIREMENTS FOR USER APPLIED OPERATIONS WERE   0.10 HOURS AT A COST OF $ 0.35 WITH AN EFFECTIVE HOURLY
LABOR RATE OF $ 3.50 PER HOUR.
REQUIRED TRANSACTIONS INCLUDE ALL EXPENSES THAT WERE REQUIRED OR NEEDED TO PERFORM ALL OF THE OPERATIONS IN THIS SIMULATION. For
example, if the user specified 100 HRS. OF LABOR AND THE SIMULATION ONLY USED 50 HR. THEN IN REQUIRED TRANSACTIONS THE LABOR
EXPENSE WILL INCLUDE COST FOR ONLY 50 HOURS OF LABOR.
SCHEDULED TRANSACTIONS INCLUDE ONLY THE EXPENSES THAT THE USER HAS SPECIFIED IN THE INPUT FORMS. THIS INCUDES DEPRECIATION.
CONTINUING THE EXAMPLE, THE USER WILL BE CHARGED FOR THE 100 HRS. OF LABOR IN SCHEDULED TRANSACTIONS REGARDLESS OF THE ACTUAL
AMOUNT OF LABOR REQUIRED TO PERFORM ALL OF THE MANAGEMENT OPERATIONS.

Figure 7 (2 of 2)

FINANCIAL STATEMENT

****

TYPE/CATEGORY/ITEM DESCRIPTION	UNITS	ID	NUMBER	$/UNIT	SUBTOTAL,$	SUBTOTAL,$	TOTAL,$	*PERCENT CHANGE FROM* START THIS YR	LAST VAL
ASSETS									
CASH									
BANK ACCOUNTS									
BANK ACCOUNTS	DOLLARS	1	0.0	1.00	0.0			-0.0	-0.0
BANK ACCOUNTS	DOLLARS	2	45966.35	1.00	45966.35			105.9	105.9
$ITEM SUBTOTAL					45966.35			105.9	105.9
$$CATEGORY SUBTOTAL						45966.35			
ACCOUNTS RECEIVABLE									
$$CATEGORY SUBTOTAL						0.0		-0.0	-0.0
NOTES RECEIVABLE									
$$CATEGORY SUBTOTAL						0.0		-0.0	-0.0
INVENTORIES									
CATTLE:FIELDS NO. 1	100 LBS	5	422.44	35.00	1485.47				
CATTLE:FIELDS NO. 1	100 LBS	7	838.86	40.00	33554.59				
CATTLE:FIELDS NO. 1	100 LBS	13	24.95	0.0	0.0				
$ITEM SUBTOTAL					35040.06		3.5	3.0	3.0
HAY	TONS	12	-425.57	65.00	-27662.09				
$$ITEM SUBTOTAL					-27662.09		-0.0	99.8	99.8
FERTILIZER	TONS	1	-43.20	167.00	-7214.39				
FERTILIZER	TONS	2	0.0	192.00	0.0				
FERTILIZER	TONS	3	0.0	151.00	0.0				
$$ITEM SUBTOTAL					-7214.39		-0.0	100.0	100.0
FUEL	GAL-LBS	1	-95.76	1.10	-105.34				
$$ITEM SUBTOTAL					-105.34		-0.0	97.9	97.9
MINERALS	LBS	1	-8216.82	0.20	-1643.36				
$$ITEM SUBTOTAL					-1643.36			99.7	99.7
$$CATEGORY SUBTOTAL						-1585.12	-104.7	-110.1	-110.1

Figure 8. Example output listing by BEEF of financial statement for a particular time period (1 of 2).

```
FIXED ASSETS
  LAND (MARKET VALUE)        1  ACRES   40.00  800.00  32000.00             0.0   0.0   0.0
  LAND (MARKET VALUE)        2  ACRES   40.00  800.00  32000.00             0.0   0.0   0.0
  LAND (MARKET VALUE)        3  ACRES   40.00  800.00  32000.00             4.8   4.8
      $$ITEM SUBTOTAL                                  96000.00
      $$CATEGORY SUBTOTAL                                        96000.00   8.1   4.8
$$$$TOTAL ASSETS                                                140381.19
********************************************************************************
LIABILITIES
  ACCOUNTS PAYABLE (1 YR OR LESS; GREATER THAN 1 YR)
      $$CATEGORY SUBTOTAL                                            0.0   -0.0  -0.0
  NOTES PAYABLE (1 YR OR LESS; GREATER THAN 1 YR)
      $$CATEGORY SUBTOTAL                                            0.0   -0.0  -0.0
  BANK ACCOUNTS              3  DOLLARS  168.36   1.00     168.36
      $$ITEM SUBTOTAL                                     168.36
      $$CATEGORY SUBTOTAL                                 168.36   -0.0   0.0   0.0
$$$$TOTAL LIABILITIES                                     168.36   -0.0   0.0   0.0
********************************************************************************
NET WORTH                                              140212.81    8.0   4.8   4.8
A "-0.0" VALUE INDICATES A ZERO IN THE DIVISOR
```

Figure 8 (2 of 2)

```
****************************************************************************************
****************************************************************************************
                              FINANCIAL RATIO ANALYSIS
TYPE/CATEGORY         FORMULA      VALUE  TREND  SYMBOL              DEFINITION                                    VALUE
****************************************************************************************
LIQUIDITY
                                                AR --ACCOUNTS RECEIVABLE:NOTES RECEIVABLE, DELAYED                  0.0
                                                     PAYMENTS, ALL ONE YEAR OR LESS.
  CURRENT            CA/CL        -0.00   0.00   CA --CURRENT ASSETS:CASH,AR,INVENTORY                          44381.23
  QUICK OR ACID TEST (CA-IV)/CL   -0.00   0.00   CL --CURRENT LIABILITIES:NOTES PYABLE, BANK DEBT                   0.0
                                                     (STATED AS ACCOUNTS PAYABLE ALL ONE YEAR OR LESS).
LEVERAGE
  DEBT TO TOTAL ASSETS TD/TA       0.00  -0.0
  TIMES INTEREST EARNED (PBT+IT)/IT -0.00  0.0   IAF--INCOME FOR FIXED CHARGES:PBT,IT,NOTES PAYABLE AND         23644.26
                                                     ACCOUNTS PAYABLE (TOTAL FOR LAST TWO ITEMS).
  FIXED COVERAGE     (AF/(IAF-PBT) -0.00   0.0   IT --INTEREST CHARGES                                             0.0
ACTIVITY
                                                IV --INVENTORY:PHYSICAL PRODUCTION AVAILABLE FOR SALE          -1585.12
                                                     EXCEPT FOR LAND, MACHINERY AND GROWING CROPS.
  INVENTORY TURNOVER  S/IV       -18.09 -11.4    NW --NEW WORTH:TOTAL ASSETS MINUS TOTAL LIABILITIES          140212.81
  AV. COLLECTION PERIOD AR/SPD     0.0  -0.0     PAT--PROFIT (OR INCOME) AFTER ALL BUT INCOME TAX             23644.26
  FIXED ASSETS TURNOVER S/FA       0.30   0.0    PBT--PROFIT (OR INCOME) BEFORE INCOME TAX                    23644.26
  TOTAL ASSETS TURNOVER S/TA       0.20   0.0    S  --SALES:INCOME EXCLUDING MISCELLANEOUS INCOME,            28670.64
                                                     THIS YEAR.
                                                SPD--SALES PER DAY:THIS YEARS SALES ONLY AND NOT
PROFITABILITY                                        INCLUDING MISCELLANEOUS INCOME.                             78.55
  PROFIT MARGIN ON SALES PAT/S     0.82   0.0    TA --TOTAL ASSETS                                           140381.19
  RETURN ON TOTAL ASSETS PAT/TA    0.17   0.0    TD --TOTAL DEBT                                                168.36
  RETURN ON NET WORTH  PAT/NW      0.17   0.0
****************************************************************************************
A ZERO VALUE INDICATES NO CHANGE WHILE A MINUS ZERO INDICATES A ZERO DIVIDE OR NOT APPLICABLE
****************************************************************************************
```

Figure 9. Example output listing by BEEF of financial ratio analysis for a particular time period.

SUMMARY OF FORAGE AND GRAIN ENTERPRISE
JAN. 1, 1981 TO JAN. 1, 1982

TYPE OF FORAGE AND GRAIN	LAND USES (ACRES)	LAND USE TOTAL LAND (%)	QUANTITY HARVESTED FORAGE IN (TONS/ACRE) GRAIN IN (BUSHELS/ACRE)	QUANTITY FED QUANTITY HARVESTED (%)	QUANTITY PURCHASED QUANTITY FED (%)	QUANTITY SOLD QUANTITY HARVESTED (%)	QUANTITY IN SOTRAGE QUANTITY HARVESTED AND PURCHASED (%)
SILAGE	0.0	0.0	0.0	0.0	0.0	0.0	0.0
HAY	0.0	0.0	0.0	0.0	99.71	0.0	0.0
HAYLAGE	0.0	0.0	0.0	0.0	0.0	0.0	0.0
CROP RESIDUE	0.0	0.0	0.0	0.0	0.0	0.0	0.0
CUBES AND PELLETS	0.0	0.0	0.0	0.0	0.0	0.0	0.0
GRAIN	0.0	0.0	0.0	0.0	0.0	0.0	0.0
	120.00	100.00					

SUMMARY OF CATTLE ENTERPRISE
JAN. 1, 1981 TO JAN. 1, 1982

LAND USE (ACRES)	LAND USE TOTAL LAND (%)	QUANTITY PRODUCED (CWT/ACRE)	QUANTITY SOLD QUANTITY PRODUCED (%)	FORAGE CONSUMED TOTAL FEED CONSUMED (%)	GRAIN CONSUMED TOTAL FEED CONSUMED (%)	PASTURE CONSUMED TOTAL FEED CONSUMED (%)	PROTEIN SUPPLEMENT CONSUMED TOTAL FEED CONSUMED (%)
120.00	100.00	3.22	111.54	39.96	0.0	60.04	0.0

Figure 10. Example output listing by BEEF of enterprise summary

93

YEARLY SUMMARY FOR THE YEAR ENDING JAN. 1, 1982

#	DESCRIPTION	CAT (1) MATURE BULLS	CAT (2) MATURE STEERS	CAT (3) YRLING BULLS	CAT (4) YRLING STEERS	CAT (5) COWS (NL-NP)	CAT (6) YRLING HEIFERS	CAT (7) COWS (P-NL)	CAT (8) COWS (L-NP)	CAT (9) COWS (L-P)	CAT (10) CALVES FEMALE	CAT (11) CALVES BULL	CAT (12) CALVES STEER	TOTALS
1	DAYS ON FARM	63.00	0.0	0.0	0.0	366.00	0.0	299.00	186.00	119.00	184.00	67.00	117.00	1401.00
2	AVERAGE NUMBER	4.40	0.0	0.0	0.0	4.13	0.0	64.63	51.11	83.33	52.21	52.21	52.21	364.24
3	AVERAGE AGE, MONTHS	25.05	0.0	0.0	0.0	77.38	0.0	77.63	77.00	79.81	3.04	1.12	4.14	45.21
4	AVERAGE WT., LB/ANIMAL	1641.96	0.0	0.0	0.0	1060.66	0.0	1111.63	1031.99	1076.16	199.97	142.56	258.49	706.28
5	NORMAL WEIGHT, LB/ANIMAL	1470.58	0.0	0.0	0.0	905.27	0.0	968.67	905.26	911.93	196.83	141.30	254.33	620.78
6	ACTUAL GAIN, LWT, LB/ANIMAL	83.04	0.0	0.0	0.0	60.56	0.0	30.98	60.55	71.25	232.68	93.03	163.65	102.13
7	NORMAL GAIN, LWT, LB/ANIMAL	7.03	0.0	0.0	0.0	0.0	0.0	95.35	0.0	19.12	230.20	90.43	186.55	94.08
	NET WEIGHT CHANGE													
8	TOTAL LB	365.36	0.0	0.0	0.0	249.86	0.0	2002.33	3094.61	5937.61	12148.56	4857.50	8544.76	37200.59
9	LB/ANIMAL-DAY	1.32	0.0	0.0	0.0	0.17	0.0	0.10	0.33	0.60	1.26	1.39	1.40	0.62
10	LB/ANIMAL-ACRE	0.69	0.0	0.0	0.0	0.50	0.0	0.26	0.50	0.59	1.94	0.78	1.36	0.85
11	LB/ACRE-DAY	0.05	0.0	0.0	0.0	0.01	0.0	0.06	0.14	0.42	0.55	0.60	0.61	0.22
12	LB/ACRE	3.04	0.0	0.0	0.0	2.08	0.0	16.69	25.79	49.48	101.24	40.48	71.21	310.00
13	LB/ANIMAL	83.04	0.0	0.0	0.0	60.56	0.0	30.98	60.55	71.25	232.68	93.03	163.65	102.13
	PURCHASE													
14	NUMBER	4.40	0.0	0.0	0.0	0.0	0.0	0.0	0.0	0.0	0.0	0.0	0.0	4.40
15	TOTAL WEIGHT, LB	7040.00	0.0	0.0	0.0	0.0	0.0	0.0	0.0	0.0	0.0	0.0	0.0	7040.00
16	AVERAGE WEIGHT, LB	1600.00	0.0	0.0	0.0	0.0	0.0	0.0	0.0	0.0	0.0	0.0	0.0	1600.00
17	TOTAL COST, $	3872.00	0.0	0.0	0.0	0.0	0.0	0.0	0.0	0.0	0.0	0.0	0.0	3872.00
18	PRICE, $/CWT	55.00	0.0	0.0	0.0	0.0	0.0	0.0	0.0	0.0	0.0	0.0	0.0	55.00
	SALES													
19	NUMBER	4.40	0.0	0.0	0.0	0.0	0.0	0.0	0.0	0.0	52.21	0.0	52.21	108.83
20	TOTAL WEIGHT, LB	7405.21	0.0	0.0	0.0	0.0	0.0	0.0	0.0	0.0	17139.67	0.0	18515.43	43060.31
21	AVERAGE WEIGHT, LB	1683.01	0.0	0.0	0.0	0.0	0.0	0.0	0.0	0.0	328.26	0.0	354.61	395.68
22	TOTAL RETURN, $	4072.87	0.0	0.0	0.0	0.0	0.0	0.0	0.0	0.0	9426.82	0.0	12035.03	25534.71
23	PRICE, $/CWT	55.00	0.0	0.0	0.0	0.0	0.0	0.0	0.0	0.0	55.00	0.0	65.00	59.30

Figure 11. Example output listing by BEEF of simulation statistics (1 of 4).

#	Item											
	MILK,LB./DAY											
24	TOTAL PRODUCED	0.	0.	0.	0.	0.	0.	17426.	0.	0.	0.	35418.
25	TOTAL DEMAND	0.	0.	0.	0.	0.	0.	17992.	68622.	21093.	57175.	146890.
26	TOTAL INTAKE	0.	0.	0.	0.	0.	0.	0.	16538.	6884.	11996.	35418.
27	TOTAL/COW	0.0	0.0	0.0	0.0	0.0	0.0	340.94	215.91	0.0	0.0	263.44
28	DEMAND/CALF	0.0	0.0	0.0	0.0	0.0	0.0	0.0	1314.30	403.97	1095.05	937.77
29	INTAKE/CALF	0.0	0.0	0.0	0.0	0.0	0.0	1.83	316.75	131.84	229.75	226.11
30	TOTAL/COW-DAY	0.0	0.0	0.0	0.0	0.0	0.0	0.0	0.0	0.0	0.0	1.82
31	DEMAND/CALF-DAY	0.0	0.0	0.0	0.0	0.0	0.0	0.0	7.14	6.03	9.36	7.64
32	INTAKE/CALF-DAY	0.0	0.0	0.0	0.0	0.0	0.0	0.0	1.72	1.97	1.96	1.84
	ME(KCAL)											
33	TOTAL INTAKE(*1.0E6)	11.36	0.0	28.29	0.0	311.33	230.38	258.59	106.61	35.10	86.53	1068.18
34	INTAKE/DAY(* 1.0E6)	0.18	0.0	0.08	0.0	1.04	1.24	2.17	0.58	0.52	0.74	6.55
35	INTAKE/ANIMAL(* 1.0E6)	2.58	0.0	6.86	0.0	4.82	4.51	3.10	2.40	0.67	1.66	2.93
36	INTAKE/ANIMAL-DAY	40976.70	0.0	18731.38	0.0	16109.73	24233.15	26077.75	11097.02	10033.97	14164.86	17877.61
37	INTAKE,ME/KG	2600.66	0.0	2231.68	0.0	2087.59	2479.57	2575.62	3429.35	3688.29	3335.60	2507.29
	PROTEIN(LB.)											
38	TOTAL INTAKE	1210.56	0.0	2761.05	0.0	29022.17	23891.38	26340.74	10738.98	3543.17	8709.07	******
39	INTAKE/DAY	19.22	0.0	7.54	0.0	97.06	128.45	221.35	58.36	52.88	74.44	659.31
40	INTAKE/ANIMAL	275.13	0.0	669.20	0.0	449.03	467.44	316.10	205.68	67.86	166.80	291.61
41	INTAKE/ANIMAL-DAY	4.37	0.0	1.83	0.0	1.50	2.51	2.66	1.12	1.01	1.43	1.78
42	INTAKE,AV. PERCENT	12.58	0.0	9.88	0.0	8.83	11.67	11.90	15.67	16.89	15.23	11.31
	DRY MATTER,LB/AN.-DAY											
43	AVAILABLE	718.38	0.0	101.05	0.0	35.46	244.15	341.53	76.70	39.06	114.78	139.24
44	QUALITY DEMAND	34.73	0.0	20.05	0.0	19.22	21.67	22.32	7.14	6.02	9.36	16.49
45	WT. DEMAND	35.05	0.0	22.51	0.0	22.51	22.51	22.32	7.14	6.03	9.36	17.79
46	INTAKE	34.73	0.0	18.50	0.0	17.01	21.54	22.32	7.13	6.00	9.36	15.72
	DRY MATTER,LB											
47	TOTAL QUALITY DEMAND	9626.	0.	30280.	0.0	371431.	205988.	221282.	68591.	21059.	57175.	985431.
48	TOTAL WEIGHT DEMAND	9716.	0.	33987.	0.	434959.	213968.	223194.	68622.	21093.	57175.	1062713.
49	TOTAL INTAKE	9626.	0.	27935.	0.	328686.	204775.	221282.	68516.	20976.	57175.	938970.

Figure 11 (2 of 4)

INTAKE,AV. LB/AN-DAY												
50 PASTURE	34.53	0.0	0.0	6.68	0.0	1.95	16.96	22.23	5.07	3.03	7.37	9.09
51 STORED	0.20	0.0	0.0	11.86	0.0	15.11	4.58	0.09	0.34	0.99	0.03	6.05
52 CREEP	0.0	0.0	0.0	0.0	0.0	0.0	0.0	0.0	0.0	0.0	0.0	0.0
53 MILK	0.0	0.0	0.0	0.0	0.0	0.0	0.0	0.0	1.72	1.97	1.96	0.59
AVAILABLE DM,AV LB/AN-D												
54 PASTURE	714.19	0.0	0.0	80.76	0.0	13.43	228.48	340.07	73.59	33.64	112.29	127.77
55 STORED	4.19	0.0	0.0	20.29	0.0	22.03	15.67	1.47	1.39	3.46	0.52	10.87
56 CREEP	0.0	0.0	0.0	0.0	0.0	0.0	0.0	0.0	0.0	0.0	0.0	0.0
57 MILK	0.0	0.0	0.0	0.0	0.0	0.0	0.0	0.0	1.72	1.97	1.96	0.59
DM SHORTAGE DAYS FROM:												
58 ME LIMITATIONS	0.0	0.0	0.0	217.00	0.0	189.00	48.00	20.00	0.0	0.0	0.0	474.00
59 DP LIMITATIONS	0.0	0.0	0.0	3.00	0.0	3.00	0.0	0.0	0.0	0.0	0.0	6.00
60 ME AND DP LIMITATIONS	0.0	0.0	0.0	3.00	0.0	3.00	0.0	0.0	0.0	0.0	0.0	6.00
61 DMD,BODY WEIGHT	63.00	0.0	0.0	366.00	0.0	299.00	186.00	119.00	184.00	67.00	117.00	1401.00
62 DM NOT AVAILABLE	0.0	0.0	0.0	83.00	0.0	80.00	3.00	0.0	3.00	3.00	0.0	172.00
INTAKE/TOTAL AVAIL.DM,%												
63 PASTURE	4.81	0.0	0.0	6.61	0.0	5.51	6.95	6.51	6.61	7.77	6.42	6.53
64 STORED	0.03	0.0	0.0	11.73	0.0	42.62	1.88	0.03	0.44	2.55	0.03	4.34
65 CREEP	0.0	0.0	0.0	0.0	0.0	0.0	0.0	0.0	0.0	0.0	0.0	0.0
66 MILK	0.0	0.0	0.0	0.0	0.0	0.0	0.0	0.0	2.24	5.04	1.71	0.43
INTAKE/QUALITY DEMAND,%												
67 PASTURE	99.42	0.0	0.0	33.32	0.0	10.17	78.26	99.61	71.06	50.40	78.69	51.13
68 STORED	0.58	0.0	0.0	59.13	0.0	78.63	21.16	0.39	4.72	16.52	0.33	36.67
69 CREEP	0.0	0.0	0.0	0.0	0.0	0.0	0.0	0.0	0.0	0.0	0.0	0.0
70 MILK	0.0	0.0	0.0	0.0	0.0	0.0	0.0	0.0	24.11	32.69	20.98	3.59
INTAKE/WEIGHT DEMAND,%												
71 PASTURE	98.51	0.0	0.0	29.69	0.0	8.68	75.34	98.76	71.03	50.32	78.69	51.13
72 STORED	0.57	0.0	0.0	52.68	0.0	67.14	20.37	0.39	4.72	16.49	0.33	34.00
73 CREEP	0.0	0.0	0.0	0.0	0.0	0.0	0.0	0.0	0.0	0.0	0.0	0.0
74 MILK	0.0	0.0	0.0	0.0	0.0	0.0	0.0	0.0	24.10	32.64	20.98	3.33

Figure 11 (3 of 4)

75	INTAKE/DM: INTAKE,% PASTURE	57.87	78.69	50.60	71.14	99.61	78.72	11.49	0.0	36.12	0.0	0.0	99.42
76	STORED	38.48	0.33	16.58	4.73	0.39	21.28	88.85	0.0	64.10	0.0	0.0	0.58
77	CREEP	0.0	0.0	0.0	0.0	0.0	0.0	0.0	0.0	0.0	0.0	0.0	0.0
78	MILK	3.77	20.98	32.82	24.14	0.0	0.0	0.0	0.0	0.0	0.0	0.0	0.0
79	INTAKE/AVAILABLE DM,%	11.29	8.15	15.35	9.30	6.53	8.82	47.96	0.0	18.31	0.0	0.0	4.83
80	INTAKE/QUALITY DMD,%	95.28	100.00	99.61	99.89	100.00	99.41	88.49	0.0	92.26	0.0	0.0	100.00
81	INTAKE/WT. DMD.,%	88.36	100.00	99.45	99.85	99.14	95.70	75.57	0.0	82.19	0.0	0.0	99.08
82	QUALITY DMD/AVAIL.DM,%	11.85	8.15	15.41	9.31	6.53	8.87	54.20	0.0	19.84	0.0	0.0	4.83
83	WT.DMD/AVAILABLE DM,%	12.77	8.15	15.44	9.31	6.59	9.22	63.47	0.0	22.27	0.0	0.0	4.88
84	QUALITY DMD/WT. DMD,%	92.73	100.00	99.84	99.96	99.14	96.27	85.39	0.0	89.09	0.0	0.0	99.08
85	NET BEEF SOLD/PROD.,%	96.83	216.69	0.0	141.08	0.0	0.0	0.0	0.0	0.0	0.0	0.0	99.96
86	NET BEEF PRODUCED,%	100.00	19.04	10.82	27.07	13.23	6.90	4.46	0.0	0.56	0.0	0.0	0.81
87	NET BEEF SALES,%	100.00	47.13	0.0	36.92	0.0	0.0	0.0	0.0	0.0	0.0	0.0	15.95
88	DM SHORTAGE,% FROM: ME LIMITATIONS	33.83	0.0	0.0	0.0	16.81	25.81	63.21	0.0	59.29	0.0	0.0	0.0
89	DP LIMITATIONS	0.43	0.0	0.0	0.0	0.0	0.0	1.00	0.0	0.82	0.0	0.0	0.0
90	ME AND DP LIMITATIONS	0.43	0.0	0.0	0.0	0.0	0.0	1.00	0.0	0.82	0.0	0.0	0.0
91	DMD,BODY WEIGHT	100.00	100.00	100.00	100.00	100.00	100.00	100.00	0.0	100.00	0.0	0.0	100.00
92	DM NOT AVAILABLE	12.28	0.0	4.48	1.63	0.0	1.61	26.76	0.0	22.68	0.0	0.0	0.0
93	NO LIMITS OF ANY TYPE	0.0	0.0	0.0	0.0	0.0	0.0	0.0	0.0	0.0	0.0	0.0	0.0

Figure 11 (4 of 4)

YEARLY SUMMARY RATIOS FOR THE YEAR ENDING JAN. 1, 1982

NO.	DESCRIPTION	VALUE	NO.	DESCRIPTION	VALUE
1	CHANGE IN NET WORTH, %	4.764	16	NET BEEF SOLD/NET BEEF PRODUCED, %	96.827
2	NET INCOME/BEEF SOLD, $/POUND	0.106	17	NET BEEF PRODUCED, LB/PASTURE CONSUMED, LB., %	6.847
3	NET INCOME/BEEF PRODUCED, $/POUND	0.101	18	NET BEEF PRODUCED, LB/STORED DM CONSUMED, LB, %	10.296
4	NET INCOME/NET BEEF SOLD, $/POUND	0.126	19	NET BEEF PRODUCED, LB/DM CONSUMED, LB, %	3.962
5	NET INCOME/ACRE, $/ACRE	37.940	20	KCAL/POUND OF BEEF PRODUCED	3897.486
6	NET INCOME/NET WORTH, % RETURN ON INVESTMENT	3.247		CONSUMPTION PER COW: *	
7	NET MARKETABLE BEEF/ACRE, POUNDS/ACRE	310.005	21	MEIKCAL:PASTURE	2237483.000
8	CONCEPTION,%	94.935	22	:STORED DM	1846376.000
9	PASTURE/TOTAL INTAKE,%	57.869	23	:TOTAL	4077663.000
10	STEER WEIGHT AT SALE, POUNDS/ANIMAL	354.614	24	DP(LB) :PASTURE	221.472
11	STEER ADJUSTED 205 DAY WEANING WEIGHT, LB/AN.	386.068	25	:STORED DM	182.760
			26	:TOTAL	403.619
	ANIMAL NUMBER PER ACRE, AVERAGE WHEN PRESENT				
12	CALVES	1.305	27	DM(LB) :PASTURE	2113.526
13	YEARLINGS AND MATURE STEERS	0.0	28	:STORED DM	1744.086
14	COWS	1.693	29	:TOTAL	3851.758
15	BULLS	0.037			

* ESTIMATED BASED ON PROPORTION OF FEED

Figure 12. Example output listing by BEEF of summary ratios for a given time period

ANALYST INPUT

The BEEF model was developed so that it might be modified by the analyst to reflect differences in geographical areas or to test the sensitivity of the system to certain input parameters. The analyst inputs reflect our best estimates of the effect of management decisions on the production process. These estimates were obtained directly and indirectly from the literature and other researchers using both inductive and deductive reasoning to obtain the best values possible.

SUMMARY

The BEEF model is the result of an interdisciplinary effort that incorporates the interactions of growing crops, grazing beef animals, energy utilization, and economics in a farm production system. It is a dynamic computer simulation that utilizes mathematical expressions in determining the physical cause-effect relationships among system components.

BEEF may be used by farmers to plan future management decisions or resource allocations. It may be used by extension specialists and teachers to demonstrate the benefits of sound management practices. It may also be used by researchers to evaluate the sensitivity of the total system to subsystem interactions and components.

The BEEF model may be used as a cropping model only. Presently, beef animals are the only livestock category that may be considered by the model user.

The agricultural system is a complex aggregate of subsystems, each somewhat dependent on the other. Maximizing the effect of one particular sector may reduce the functional capability of the total system. What appears to be a sound management practice may result in reduced production for reasons not readily apparent until the system component interactions are evaluated. When this "counter-intuitive" behavior (Forrester, 1971) is analyzed, it leads to greater understanding of the system by the farmer, extension specialist, and researcher. The BEEF model was developed to aid in this understanding.

REFERENCES

Bridges, T.C., and E.M. Smith. 1979. "A Method for Determining the Total Energy Input for Agricultural Practices." Transactions of the ASAE. 22(4): 781-784.

Forrester, Jay W. 1968. Principles of Systems, 2nd Ed., Wright Allen Press, Cambridge, MA.

_____. 1971. "Counter-intuitive Behavior of Social Systems." Simulation February, pp. 61-76.

Loewer, O.J., E.M. Smith, G. Benock, N. Gay, T. Bridges and L. Wells. 1980. "Dynamic Simulation of Animal Growth and Reproduction." Transactions of the ASAE. 23(1):131-138.

Loewer, O.J., E.M. Smith, G. Benock, T. Bridges, L.G. Wells, N. Gay, S. Burgess, L. Springate, and D. Debertin. 1981. "A Simulation Model for Assessing Alternative Strategies of Beef Production with Land, Energy and Economic Constraints." Transactions of ASAE, 24(1):164-173.

Loewer, O.J., E.M. Smith, N. Gay, and R. Fehr. 1983a. "Incorporation of Environment and Feed Quality into a Net Energy Model for Beef Cattle." Agricultural Systems. 11:67-94.

Loewer, O.J., E.M. Smith, K.L. Taul, L.W. Turner, and N. Gay. 1983b. "A Body Composition Model for Predicting Beef Animal Growth." Agricultural Systems. 10:245-256.

Loewer, Otto J., K.L. Taul, L.W. Turner, N. Gay, and R. Muntifering. 1985. "Modeling of Selective Grazing by Beef Animals as Influenced by Environment." ASAE Paper No. 85-4008. St. Joseph, MI.

Lofgreen, G.P., and W.N. Garrett. 1968. "A System for Expressing Net Energy Requirements and Feed Values for Growing and Finishing Beef Cattle " J. of Animal Science. 27:793-806.

National Research Council Committee on Animal Nutrition. 1969. United States-Canadian tables on Feed Composition.

Pritsker, A. Allan B. 1974. The GASP IV Simulation Language. John Wiley and Sons, New York.

Smith, E.M., L. Tharel, M.A. Brown, C.T. Dougherty, and K. Limbach. 1985. "A Simulation Model for Managing Perennial Grass Pastures. I. Structure of the Model." Agricultural Systems. 17:155-180.

Smith, E.M., and O.J. Loewer. 1983. "Mathematical-logic to Simulate the Growth of Two Perennial Grasses." Transactions of the ASAE, 26(3):878-883.

Taul, Keith L. 1984. "Modeling the Surface Area of Cattle as a Function of Body Dimensions and Geometric Shapes." Unpublished M.S. Thesis. Department of Agricultural Engineering, University of Kentucky, Lexington, KY.

Taul, K.L., O.J. Loewer, L.W. Turner, N. Gay, and R. Muntifering. 1985. "Modeling Surface Area of Beef Steers." ASAE Paper No. 85-4009, St. Joseph, MI.

Turner, Larry W. 1984. "Modeling Thermo Regulation and Heat Transfer within the Beef Animal." Ph.D. Dissertation, Department of Agricultural Engineering, University of Kentucky, Lexington, KY.

Walker, J.N., E.M. Smith, O.J. Loewer, Jr., G.T. Benock, N. Gay, S. Burgess, L.G. Wells, T.C. Bridges, L. Springate, J. Boling, G. Bradford, and D. Debertin. 1977a. "BEEF: Beef, Energy and Economic Evaluator for Farms, Vol. I - Discussion and Results of the Study." Final Report of the Study to the National Science Foundation.

_____. 1977b. "BEEF: Beef, Energy and Economic Evaluator for Farms, Vol. II - Discussion and Listing of the Model." Final Report of the Study to the National Science Foundation.

_____. 1977c. "BEEF: Beef, Energy and Economic Evaluator for Farms, Vol. III - Input Forms for the Model." Final Report of the Study to the National Science Foundation.

Watson, Vance W., and Chester M. Wells, (Eds.) 1985. "Simulation of Forage and Beef Production in the Southern Region." Southern Cooperative Series Bulletin 308. Mississippi Agricultural and Forestry Experiment Station, Mississippi State University, Mississippi State, MS.

V

A Discussion of the Dynamic Simulation Model "PASTURE"

*L.M. Tharel, M.A. Brown, E.M. Smith,
and C.T. Dougherty*

INTRODUCTION

The pasture is the land unit which is most often associated with beef production systems, and most beef animals begin and spend a good part of their lives on the pasture. Forage crops, particularly perennial grasses, are basic elements of pastures and supply much of the feed for beef animals. These perennial grasses also stabilize the land unit by protecting the soil from water and wind erosion.

The management of these pastures and the beef animals which utilize the feed produced by the pastures is an important issue in planning and executing beef production systems. Research should provide the fundamental knowledge which can be used as a basis for developing management strategies which have the highest probability of success in particular situations.

There are literally as many production systems as there are beef producers, and for each of these systems there are many management strategies which need to be evaluated in order that producers in each situation can effectively utilize resources to produce feed for animals. The use of traditional field experiments to evaluate a large number of management strategies simultaneously for many different production situations is impractical, because such experiments would require an enormous commitment of money, land, and scientific man-years (Smith, et. al., 1984).

USDA, ARS, Booneville, AR; Agricultural Engineering Department, University of Kentucky, Lexington; Agronomy Department, University of Kentucky, Lexington.

*Parts of this manuscript including tables are taken from Smith, et al, 1984 and 1985.

Grazing systems research may be conducted with (1) whole systems; (2) relevant parts of systems; (3) component processes or relationships leading to the construction of theoretical models; and, (4) surveys or other observational studies (Spedding, 1975). Simulation models, based upon fundamental biological and physical principles which have been experimentally validated, can be used to conduct extensive evaluations at relatively low costs. In fact with present computer technology, reliance on simulation models to evaluate many different management scenarios before implementation is common in many industries, especially management strategies which would affect the future viability of the company. Agriculture should be at the forefront in the use of this technology.

A number of forage-beef simulation models have been developed to address the need for the ability to evaluate production alternatives. Mathematically logical functions (sub-models) that describe crop growth and quality as influenced by genetic potential, air temperature, day length, rainfall, leaf area, nutrients (N, P, K), and harvesting strategy have been developed and validated (Smith, 1983; Brown, 1982; Neels, 1981; Ewen, 1980; Bingner, 1980). These functions have been programmed into a crop-growth model that is structured to simulate the effects of different management options on the performance of perennial grass pastures (Smith, et. al., 1985). There is a need for additional work in developing logic and program technology to simulate forage production in a manner more closely associated with actual biological response of the crops particularly with forages adapted to southern cilmates. The purpose of this work is to develop the logic of functional relationships of crop response to genetic potential and environment and to convert this logic into usable machine code to simulate growth for perennial forage grasses.

"PASTURE" is a dynamic simulation model which can be used to determine the quantity, quality, and seasonal distribution of the feed supply for beef animals. It reflects the effects of plant genetic potential, environment, and management on these attributes of the feed supply.

A growing crop is a dynamic phenomenon which functions as a self-sustaining mechanism responding to its environment. The leaves of the growing crop receive solar energy and photosynthetically manufacture carbohydrates which form the resource base for growth. This resource base is used along with nutrients and water from the soil and air to produce dry matter, which is a manifestation of growth. Growth is dependent upon the genetic potential of the crop, environment, and management. Growth and environment are dynamic phenomena and management can be both dynamic and discrete.

A dynamic simulation model must be based upon mathematical-logical relationships which follow the changes in selected attributes of an entity as time progresses. The mathematical-logical relationships can be discrete or quasi-continuous algorithms of specific data or continuous equations based upon fundamental physiology of growth and physics of the environment. Continuous equations are generally considered better than alogrithms for simulating dynamic systems.

The nonspecific crop growth model is based upon mathematical continuous equations, with time as the independent variable. These equations are abridgements of many complex physical and chemical processes which occur in the growth of a crop based on contemporary knowledge available in the literature. The validity of the hypotheses upon which these equations are based has been tested, but the overall system of equations is subject to validation based on biological research and use as would be any technology developed.

Many models are site specific, crop specific, and/or management specific, because their bases are algorithms of site specific, crop specific, and/or management specific data. The usefulness of such models is restricted to the physiographic areas, crops, and management regimes from which the specific data base was derived.

A nonspecific model is not restricted by site, crop, or management specificity and, consequently, can be used in different physiographic areas by specifying those parameters which relate site, crop, and management variables to the attributes being simulated. The nonspecificity is accomplished by developing continuous mathematical-logical equations which describe the fundamental relationships among the variables that affect time-related changes in the attributes being simulated by the model. These mathematical-logical equations form the basis for the nonspecific model, as opposed to the algorithms of specific data which form the bases for specific models (Smith and Loewer, 1981).

PASTURE MODEL APPLICATIONS

Significant problems and sub-problems associated with pastures may be evaluated with the model. The PASTURE model can be a useful analytical tool in dealing with these problems.

Under-Utilization of Forage

This problem is widespread and is the root cause of other problems, such as: encroachment of weeds and brush into pastures, loss of legumes from mixed species, and poor quality forage during much of the grazing and haying seasons. Under-utilization also tends to inhibit the transfer of technology because it obliterates the beneficial effects that new technology offers.

Increasing stocking rates would seem to solve this problem, however, producers who adopt this solution incur management responsibilities that are not usually practiced when under-utilization is prevalent. Relevant research may be conducted that provides a knowledge base for delineating those management responsibilities that increased stocking rates would incur. The PASTURE model provides a technological tool whereby different scenarios for managing pastures to improve forage utilization could be simulated. Simulated results could form bases for planning future research, farm demonstrations, and developing viable management options.

The South has the forage base and concomitant climate that provide a competitive edge over many other areas for grazing beef cattle. Improved forage utilization combined with this competitive edge could support an increase in animal numbers and, consequently, agricultural income.

Uneven Distribution of Forage

This is a persistent problem with pastures and is not likely to be eradicated. In fact, this inherent characteristic of forage growth probably should be considered as a blessing instead of a problem, because it provides a reliable source of stored feed that can be used to supplement pastures during expected or unexpected periods of low forage availability.

Solutions to uneven distribution lie primarily in the delineation of grazing and haying management that effectively utilizes the seasonal production characteristics of forages. Effective utilization depends upon the objectives of the beef production program, because the performance of beef animals is directly related to what they consume.

There are many beef production objectives and forages that can be utilized to achieve these objectives. The PASTURE model could be used to simulate many different scenarios, and the results would help identify grazing and haying management that effectively utilizes different forages for different beef production objectives. The stochastic nature of the model would be especially relevant because seasonal production is so dependent upon weather.

Fertilization

The general proposition that most pastures exhibit low productivity because they are not fertilized could be documented with statistics. However, this problem is usually masked by the problem of under utilization. Producers, evidently, fail to see the benefits that might accrue as a result of fertilizing pastures, because they are not effectively utilizing their present pastures.

Fertilization programs for pastures should be developed as an integral feature of an overall management plan for utilizing pastures. The quantity applied and the time of application should be designed to give the pasture the nutrients that it needs to produce the quantity and quality of forage that are required in the utilization strategy. For example, if a rotation grazing system includes several pasture fields and each field is allowed to rest (forage regrowth) for 30 days, fertilizer might be applied at the beginning of the rest period (end of a grazing period) to supply nutrients that the forage needs for regrowth.

The PASTURE model simulates the uptake of primary nutrients (N, P, K) by growing forage and loss of N due to leaching. It reflects the interactions with weather, physiological age of the forage, and management in simulating the uptake and loss of nutrients. In view of these interactions, the stochastic nature of the model is appropriate for simulating different scenarios for fertilizing pastures as an integral feature of grazing and haying management.

Drought Periods

Drought periods represent significant problems in pastures, because they occur at various times during every annual cycle. The time when drought periods occur and their duration determine the level of significance of the problem. With cool-season forages, drought periods that occur during spring and fall seasons are more serious than those during summer or winter when it might be too hot or cold for adequate forage growth. Drought periods of long duration are serious at any time, because the soil moisture is depleted to such an extent that a considerable amount of rain is required just to replenish the soil moisture to the condition where the forage can efficiently utilize it for growth.

Irrigation seems to be an obvious solution to this problem, however supplemental feeding of hay might also be a viable management option during drought periods. The PASTURE model simulates the interactions among rainfall, physiological age of the forage, photosynthetically active leaf area, air temperature, fertility, and management, consequently it can monitor the condition of growing forage with respect to moisture stress.

The stochastic nature of the model is appropriate for long term simulation of different grazing management scenarios to determine the frequency of occurrence of drought periods. Then, these scenarios can be simulated with irrigation moisture applied during these drought periods to determine whether or not irrigation is an effective management option in grazing management.

Forage managers and researchers should respond to problems that exist in pastures by making management decisions and/or recommendations to producers that enable them to solve problems. Pastures, unlike other enterprises, involve daily harvesting (grazing) of forage which can dramatically affect the growth of the forage. Grazing management must be a feature of every recommendation concerning pastures.

A typical response to a pasture problem is to recommend a forage that seems to have characteristics applicable to the problem. Such a recommendation may or may not be effective depending upon subsequent grazing management. In many cases, grazing management is the root cause of the problem, not the forage. Therefore, grazing management should be an integral feature of each problem analysis, because management interacts so strongly with forage, animal, and weather variables.

The objective of PASTURE was to structure a model specifically to enable users and analysts to evaluate stochastically management options which include number of fields in a grazing system, rotation period, and grazing rate.

Stochastic simulations of many different scenarios can generate simulated data that can be statistically analyzed. This procedure allows statistical probability to be a basis for recommending possible specific solutions from many possible solutions. These applications are minimal in scope. The possible utilization of the model in problem solving is limited to the imagination and practical application of the user.

MODEL STRUCTURE

The model is structured specifically to enable users to evaluate management options stochastically. The random variables in the model may be weather as years or blocks of years, or crop and/or field variables which may be generated stochastically to simulate field replicates. The model was built around a valid crop-growth model (Smith and Loewer, 1983) for perennial grasses.

Definition of Terms

Run
: This is a group of consecutive days. The model output provides data that character-ize each run.

Simulation
: This is a complete execution of the model for one set of input data. A simulation consists of a number of consecutive runs.

Fields
: Fields indicate the number of fields to be used in a grazing system.

Rotation Period
: This is the number of consecutive days grazing on each consecutive field. For example, if there are three fields and the rotation period is five days, Field 1 will be grazed for five days, followed by Field 2 for five days, followed by Field 3 for five days, etc.

Simulated Grazing Rate
: This variable refers to the amount of dry matter which is specified for harvest. This variable is difficult to study in actual grazing experiments because of confounding by the variability amoung the animals which are used to obtain different grazing rates. However, the model can simulate different specified grazing rates by simulating different daily dry matter harvests in kilograms per hectare per day; thus, confounding is avoided.

Harvest Rate
: Harvest rate is the amount of dry matter actually harvested. The harvest rate may be less than the simulated grazing rate if the dry matter is not available to be harvested.

The model structure consists of two models: (1) a main program and (2) a subroutine GROW. A brief description of the basic program logic of each of these models follows. This is a general discussion and does not include the detailed mathematically logical rela-tionships that characterize the model. A more detailed discussion of the model structure that includes these relationships may be obtained from other publications (Smith, et al., 1984, 1985).

Main Program

The main program contains read statements and a subroutine (RDATA) for those inputs which are not field-dependent and controls the flow of information and the order in which calculations are performed during the course of a simulation. The operations controlled by the main program include (1) read and store input data that are the same for all fields in a simulation; (2) call subroutine GROW once for each field in a simulation; (3) prepare output data and store these data in an encoded array; and (4) write the output data onto a disk dataset.

Input Data

The input data that are read and stored for subsequent use in the course of a simulation are as follows:

1. The number of fields to be used in a grazing system, the number of runs per year, and the number of years spanned by a simulation.
2. The initial conditions for a simulation. Table 1 gives a description for each of these variables.
3. The normal monthly rainfall values for the simulation site.
4. The genetic variables that define the crop to be used in the simulation. This program assumes that the same crop will be used on all fields, but a projected revision to the model will enable the utilization of multiple crops. Table 2 gives a description of these variables. Table 3 suggests ranges for crop variables for different grasses and legumes which were taken from the literature. These values may be changed or refined as users feel necessary. Suggested ranges for the following crops have also been developed; timothy, smooth bromegrass, Midland bermudagrass, bahiagrass, carpetgrass, johnsongrass, Pangola digitgrass, and Birdsfoot trefoil.
5. The beginning date (month, day, year) of the initial run and the ending dates of all runs. A subroutine RDATA accepts these input data and calculates the beginning date for each consecutive run. These data are converted to a Julian format and stored in an array.
6. The encoding for the output data array that consists of a replicated code number, a number-of-field code number, a rotation period code number, and a grazing-rate code number.

Output Data

The main program calls subroutine GROW, which simulates all activities on each field for the entire simulation period and returns 33 data arrays. These are then combined into one final data array. The values in the final data array are encoded by reading the

Table 1. Variables that Describe the Initial Conditions for a Simulation.

Variable	Description
IMMAX	The current month.
IDDMAX	The current day of the month.
IYYMAX	The current year.
NYEAR	The first year for which weather data is needed in the simulation. (Weather data will be needed (M) days before the simulation begins, where (M) is the maximum value used for PERIOD (see Table 5).)
NYEAR1	The last year for which weather data is needed.
ISTATE[a]	The state identification number.
ISTAT[a]	Station identification number.
IFL[a]	Input device code. A Fortran file that temporarily stores weather data for the program's use.
XLAT	The lattitude of the station, in degrees.
RAINP1	Weight given to current conditions when finding the rainfall factor. This number is a fraction (0 < RAINP1 < 1.0) and is about 0.75 for perennial grasses (see Table 3).
NRUNS	The total number of runs in the simulation. This number is currently 80.
ZZZ	Daylength below which a crop does not utilize stored non-structural carbohydrates to support growth, in hours (Table 3).

[a]These input variables are local to the University of Kentucky Computer Center. Users in other locations will need to develop their own method of supplying weather data.

Table 2. Variables[a] that Describe the Genetic
Potential of the Crop to be Simulated.

Variable	Description
XM	The minimum air temperature at which the crop will grow (Smith and Loewer, 1983).
XN	The fraction of the cell content that is nitrogen (Burton et al., 1969).
R	Optimum rate of production of non-structural carbohydrates that occurs when the temperature is optimum (Kg/ha/hr or Lb/ac/hr) (Smith and Loewer, 1983).
QF	The lethal low temperature at which the crop dies (Sallsbury, 1969).
QH	The lethal high temperature at which the crop dies.
QR	Optimal temperature for crop growth.
XMM	Maximum air temperature at which crop will grow.
QQ1	The amount of accumulated dry matter in a crop that provides enough leaf area for maximum growth rate (Kg/ha or Lb/ac) (Brown et al., 1968).
SAXBDM	Maximum quantity of dry matter in the non-harvestable stored reserves partition, XBDM (Kg/ha or Lb/ac) (Smith et al., 1984).
CWALL1	Fraction of one day old new growth that is cell wall (Burton et al., 1969).
CWALL2	Fraction of new partitiion that is cell wall when it is XLIFE days old.
CWALL3	Fraction of old partition that is a cell wall when it is XDEAD days old.
CWALL4	Fraction of dead material that is cell wall.
DIG1	Fraction of dry matter in very young growth that is digestible.

Continued

Table 2. Variables[a] that Describe the Genetic Potential
of the Crop to be Simulated (continued).

Variable	Description
DIG2	Fraction of dry matter in senescent crop material that is digestible.
DIG3	Minimum fraction of dry matter in dead crop material that is digestible.
XP	Fraction of cell content that is phosphorus.
XK	Fraction of cell content that is potassium.
HALF	Half-life of the non-structural carbohydrates stored in the dead partition in days.
XSTORM	Fraction of the crop material in the new and old partitions that can be stored non-structural carbohydrates.
REMOVE	Fraction of SAXBDM that can be removed to support the crop growth (Davidson, 1964).
XLIFE	The physiological age at which material passes from the new to the old partition. This is the number of days required for a crop leaf to reach its mature size.
XLIFED	The daylength used to estimate XLIFE in hours. This is the average daylength during the period of days required for a leaf to reach its mature size (Murata, 1965).
XDEAD	Physiological age at which material passes from the old to the dead partition. This is the number of days required for a crop leaf to grow from inception to complete senescence.
XLEAF	Lowest possible value of leaf area factor. This is the fraction of the optimum growth rate (R) that will occur when there is no photosynthetically active leaf area. Default value is .001.

[a]Temperatures should be expressed in °C or °F, and all fractions should be greater than or equal to zero and less than or equal to one.

Table 3. Suggested Values for Variables of the Genetic Potential for Crops to be Simulated.

Crop Variable	Bluegrass	Fescue	Orchardgrass	Coastal Bermudagrass
XM (°F)	32.0 – 38.0	38.0 – 47.0	36.0 – 45.0	67.0 – 70.0
QR (°F)	55.0 – 60.0	62.0 – 67.0	60.0 – 65.0	82.5 – 85.5
XMM (°F)	78.0 – 82.0	86.0 – 94.0	84.0 – 90.0	108.0 – 112.0
R (lb/ac/da)	6.5 – 7.5	7.2 – 10.8	7.2 – 10.8	10.0 – 11.5
QF (°F)	3.0 – 7.0	9.5 – 10.5	8.5 – 9.5	38.0 – 42.0
QH (°F)	92.0 – 96.0	95.0 – 100.0	93.0 – 98.0	125.0
QQ1 (lb/ac)	800.0 – 1000.0	900.0 – 1400.0	900.0 – 1200.0	800.0 – 1000.0
XLEAF	0.001	0.001	0.001	0.001
SAXBDM (lb/ac)	300.0 – 400.0	320.0 – 480.0	400.0 – 500.0	500.0 – 600.0
CWALL1	0.52 – 0.56	0.58 – 0.60	0.56 – 0.58	0.68 – 0.70
CWALL2	0.58 – 0.60	0.61 – 0.63	0.60 – 0.62	0.71 – 0.72
CWALL3	0.62 – 0.64	0.64 – 0.66	0.63 – 0.65	0.73 – 0.77
CWALL4	0.66 – 0.68	0.67 – 0.71	0.67 – 0.70	0.77 – 0.79
DIG1	0.66 – 0.70	0.64 – 0.66	0.65 – 0.69	0.54 – 0.59
DIG2	0.61 – 0.65	0.59 – 0.61	0.60 – 0.62	0.50 – 0.54
DIG3	0.56 – 0.60	0.52 – 0.56	0.54 – 0.58	0.35 – 0.38
XN	0.050 – 0.055	0.052 – 0.057	0.057 – 0.059	0.07 – 0.072
XP	0.010 – 0.014	0.013 – 0.017	0.015 – 0.019	0.012 – 0.015
XK	0.040 – 0.044	0.042 – 0.048	0.045 – 0.050	0.049 – 0.053
XSTORM	0.25 – 0.35	0.25 – 0.35	0.25 – 0.35	0.17 – 0.20
REMOVE	0.45 – 0.55	0.45 – 0.55	0.45 – 0.55	0.45 – 0.55
ZZZ (hr)	12.0	12.0	12.0	12.0
HALF (da)	76.0 – 84.0	81.0 – 99.0	80.0 – 88.0	85.0 – 90.0
HALFS (da)	36.0 – 54.0	40.0 – 50.0	38.0 – 48.0	40.0 – 50.0
XLIFE (da)	28	28	28	28
XLIFED (hr)	12.0	12.0	12.0	12.0
XDEAD (da)	56	56	56	56
XLEG	.FALSE.	.FALSE.	.FALSE.	.FALSE.
RAINP1	0.74 – 0.78	0.70 – 0.80	0.67 – 0.74	0.72 – 0.75

Parameter				
XM (°F)	66.0 — 69.0	40.0 — 45.0	55.0 — 58.0	50.0 — 55.0
QR (°F)	83.0 — 87.0	70.0 — 75.0	68.0 — 72.0	65.0 — 72.0
XMM (°F)	108.0 — 112.0	90.0 — 95.0	90.0 — 93.0	85.0 — 90.0
R (lb/ac/da)	9.0 — 11.0	7.0 — 8.0	9.5 — 11.5	8.5 — 9.5
QF (°F)	38.0 — 42.0	20.0 — 22.0	2.0 — 4.0	5.0 — 7.5
QH (°F)	125.0	108.0 — 110.0	110.0 — 115.0	98.0 — 102.0
QQ1 (lb/ac)	900.0 — 1100.0	800.0 — 1000.0	900.0 — 1100.0	850.0 — 950.0
XLEAF	0.001 — 0.001	0.001 — 0.001	0.001	0.001
SAXBDM (lb/ac)	500.0 — 600.0	300.0 — 400.0	500.0 — 600.0	450.0 — 550.0
CWALL1	0.69 — 0.71	0.62 — 0.65	0.38 — 0.40	0.39 — 0.41
CWALL2	0.71 — 0.72	0.66 — 0.68	0.40 — 0.44	0.42 — 0.44
CWALL3	0.73 — 0.77	0.72 — 0.75	0.46 — 0.48	0.46 — 0.48
CWALL4	0.77 — 0.79	0.79 — 0.81	0.52 — 0.56	0.54 — 0.58
DIG1	0.53 — 0.58	0.56 — 0.60	0.65 — 0.68	0.64 — 0.69
DIG2	0.48 — 0.52	0.45 — 0.47	0.63 — 0.65	0.62 — 0.63
DIG3	0.30 — 0.34	0.43 — 0.46	0.58 — 0.60	0.54 — 0.56
XN	0.069 — 0.071	0.055 — 0.059	0.09 — 0.12	0.075 — 0.085
XP	0.011 — 0.013	0.012 — 0.015	0.018 — 0.02	0.016 — 0.017
XK	0.048 — 0.052	0.044 — 0.048	0.06 — 0.07	0.052 — 0.06
XSTORM	0.15 — 0.19	0.18 — 0.21	0.25 — 0.30	0.25 — 0.30
REMOVE	0.45 — 0.55	0.35 — 0.45	0.40 — 0.55	0.40 — 0.50
ZZZ (hr)	12.0 — 12.0	12.0 — 12.0	12.0 — 12.0	12.0
HALF (da)	85.0 — 90.0	85.0 — 90.0	75.0 — 80.0	75.0 — 80.0
HALFS (da)	40.0 — 50.0	40.0 — 50.0	40.0 — 50.0	40.0 — 50.0
XLIFE (da)	28	28	28	28
XLIFED (hr)	12.0	12.0	12.0	12.0
XDEAD (da)	56	56	56	56
XLEG	.FALSE. — .FALSE.	.FALSE. — .FALSE.	.TRUE. — .TRUE.	.TRUE. — .TRUE.
RAINP1	0.72 — 0.75	0.60 — 0.65	0.50 — 0.60	0.55 — 0.65

replication code number (REP), number of fields code number (FLDNUM), rotation-period code number (ROTATP), and grazing rate code number (GR). This encoded final data array is then written onto a disk data set with a write statement. Thirty-three different attributes are included in the final data array. Values for each of these attributes (Table 4) are stored on the disk data set for each simulation.

Appropriate statistical programs (SAS, 1979) can read these encoded data values from the disk data set and perform desired analyses of the data.

Subroutine Grow

The main program calls this subroutine once for each field included in a simulation. The subroutine combines the field-independent inputs passed from the main program with inputs that are read within to determine the data values for each of the 33 attributres to be included in the final data array. The data are stored in 33 arrays that are returned to the main program where they are combined to produce a final array, which is written onto a disk data set.

Table 4. Location and Description of 33 Output
Attributes from RRth Run Yth Year.

Location in Array Final	Description
(1,RR,Y)	The percentage of days during each run when XM < QMIN < XMM and QMAX < XMM. QMIN and QMAX are respectively, minimum and maximum air temperatures.
(2,RR,Y)	The percentage of days during each run when QMAX < XM.
(3,RR,Y)	The percentage of days during each run when QMIN < XM and QMAX < XMM.
(4,RR,Y)	The percentage of days during each run when QR < QMIN.
(5,RR,Y)	The percentage of days during each run when QMAX < QR.

Continued

Table 4. Location and Description of 33 Output
Attributes from RRth Run Yth Year
(continued).

Location in Array Final	Description
(6,RR,Y)	The ratio of actual/normal rainfall during each run.
(7,RR,Y)	The percentage of days during each run when there was no moisture stress.
(8,RR,Y)	The percentage of days during each run when there was moderate moisture stress.
(9,RR,Y)	The percentage of days during each run when there was severe moisture stress.
(10,RR,Y)	The quantity of available dry matter at the beginning of each run.
(31,RR,Y)	The quantity of available dry matter remaining at the end of each run.
(13,RR,Y)	The total quantity of dry matter harvested during each run.
(11,RR,Y)	The number of daily harvests during each run.
(12,RR,Y)	The average quantity of dry matter harvested during each daily harvest.
(14,RR,Y)	The percentage of the harvested dry matter which came from new growth.
(15,RR,Y)	The percentage of the harvested dry matter which came from old growth.
(16,RR,Y)	The percentage of the harvested dry matter which came from dead growth.
(17,RR,Y)	The cell wall content of the harvested dry matter in percent.
(18,RR,Y)	The digestible cell wall content of the harvested dry matter in percent.
(19,RR,Y)	The digestible dry matter content of the harvested dry matter in percent.

Continued

Table 4.   Location and Description of 33 Output
           Attributes from RRth Run Yth Year
           (continued).

Location in Array Final	Description
(20,RR,Y)	The total nitrogen content of the harvested dry matter in percent.
(21,RR,Y)	The percentage of the harvested dry matter which is nitrogen that is contained in the cell content (digestible nitrogen).
(22,RR,Y)	The phosphorus content of the harvested dry matter in percent.
(23,RR,Y)	The potassium content of the harvested dry matter in percent.
(24,RR,Y)	The quantity of nitrogen extracted from the soil during each run.
(25,RR,Y)	The quantity of nitrogen lost from the soil due to leaching during each run.
(26,RR,Y)	The quantity of phosphorus extracted from the soil during each run.
(27,RR,Y)	The quantity of potassium extracted from the soil during each run.
(28,RR,Y)	The percentage of days during each run when growth was limited by availability of nitrogen.
(29,RR,Y)	The percentage of days during each run when growth was limitied by availability of phosphorus.
(30,RR,Y)	The percentage of days during each run when growth was limitied by availability of potassium.
(32,RR,Y)	The quantity of dry matter remaining in the non-harvestable reserve partition of the crop (XBDM) at the end of each run.
(33,RR,Y)	The number of days in each run.

Input Data

The input data that are read and stored within this subroutine describe each field with respect to initial conditions, weather, scheduled harvesting strategy and scheduled applications of fertilizer (N, P, K). The input data that describe each field with respect to initial conditions are given in Table 5. These data have to be supplied for each field that is included in a simulation, and they can be different for each field. Table 6 is a list of suggested ranges for initial crop conditions. Ranges have been developed for the additional crops previously mentioned. Tables 7 and 8 give ranges for initial soil nutrient and soil moisture conditions.

Actual weather data from some historical period for the simulation site are used as input data. These data are read from a temporary Fortran file that is created by a separate program. Five subroutines are used to read the data from this temporary file and store the data in arrays that will subsequently be used in relationships that determine crop growth.

Table 5. Variables that Describe the Initial Conditions for each Field in a Simulation.

Variable	Description
PERIOD	The number of days that rainfall can be stored and used by the crop. This number is about fifteen and can be no greater than thirty.
XBDM	Initial amount of dry matter in the non-harvestable stored reserve partition (Kg/ha or Lb/ac).
XNGROW	Initial quantity of dry matter in the new partition (Kg/ha or Lb/ac).
XNGONT	Initial physiological age of the new partition in days.
XDGROW	Initial quantity of dry matter in the dead partition (Kg/ha or Lb/ac).
XDGONT	Initial physiological age of the dead partition in days.

Continued

Table 5. Variables that Describe the Initial Conditions for each Field in a Simulation (continued).

Variable	Description
XSTORN	Initial quantity of non-structural carbohydrates stored in the new partition (Kg/ha of Lb/ac).
XSTORO	Initial quantity of non-structural carbohydrates stored in the old partition (Kg/ha or Lb/ac).
XSTORD	Initial quantity of non-structural carbohydrates stored in the dead partition (Kg/ha or Lb/ac).
XNI	The total amount of nitrogen contained in the soil (Kg/ha or Lb/ac).
XPI	The total amount of phosphorus contained in the soil (Kg/ha or Lb/ac).
KKI	The total amount of potassium contained in the soil (Kg/ha or Lb/ac).
RKST	The amount of potassium stored in the plant (Kg/ha or Lb/ac).
XNIM	Minimum quantity of nitrogen always in the soil (Kg/ha or Lb/ac).
XPIM	Minimum quantity of phosphorus always in the soil (Kg/ha or Lb/ac).
XKIM	Minimum quantiy of potassium always in the soil (Kg/ha or Lb/ac).
XNLT	Fraction of nitrogen in the soil that is unavailable for immediate crop growth.
XPLT	Fraction of phosphorus in the soil that is unavailable for immediate crop growth.
XKLT	Fraction of potassium in the soil which is unavailable for immediate crop growth.
HALFN	Half-life of the nitrogen-leaching process in days.

Table 6. Suggested Values for Variables of Initial Crop Conditions for Grasses and Legumes.

Crop Variable	Bluegrass	Fescue	Orchardgrass	Coastal Bermudagrass
XBDM (lb/ac)	300.0 - 400.0	320.0 - 480.0	400.0 - 500.0	500.0 - 600.0
XNGROW (lb/ac)	100.0 - 150.0	0.0	0.0	0.0
XNGONT (days)	10.0 - 12.0	0.0	0.0	0.0
XDGROW (lb/ac)	0.0	200.0 - 250.0	200.0 - 250.0	0.0
XDGONT (days)	0.0	30.0 - 35.0	30.0 - 35.0	0.0
XDGROW (lb/ac)	200.0 - 250.0	400.0 - 500.0	400.0 - 500.0	500.0 - 600.0
XDGONT (days)	60.0 - 65.0	60.0 - 65.0	60.0 - 65.0	90.0 - 95.0
XSTORN (lb/ac)	30.0 - 45.0	0.0	0.0	0.0
XSTORO (lb/ac)	0.0	60.0 - 75.0	60.0 - 75.0	0.0
XSTORD (lb/ac)	20.0 - 25.0	40.0 - 50.0	40.0 - 50.0	30.0 - 40.0
RKST (lb/ac)	0.0	0.0	0.0	0.0

Crop Variable	Commom Bermudagrass	Dallisgrass	Alfalfa	Red Clover
XBDM (lb/ac)	500.0 - 600.0	300.0 - 400.0	500.0 - 600.0	450.0 - 550.0
XNGROW (lb/ac)	0.0	100.0 - 150.0	0.0	0.0
XNGONT (days)	0.0	10.0 - 12.0	0.0	0.0
XDGROW (lb/ac)	0.0	0.0	0.0	0.0
XDGONT (days)	0.0	0.0	0.0	0.0
XDGROW (lb/ac)	500.0 - 600.0	200.0 - 300.0	0.0	0.0
XDGONT (days)	90.0 - 95.0	60.0 - 65.0	0.0	0.0
XSTORN (lb/ac)	0.0	20.0 - 30.0	0.0	0.0
XSTORO (lb/ac)	0.0	0.0	0.0	0.0
XSTORD (lb/ac)	20.0 - 25.0	10.0 - 15.0	0.0	0.0
RKST (lb/ac)	0.0	0.0	0.0	0.0

Table 7. Suggested Values for N,P,K and (XNIM, XPIM, and XKIM) for Four Different Soil Conditions.

Soil Condition	XNIM (lb/ac)	XPIM (lb/ac)	XKIM (lb/ac)
High P - High K	0.40 - 0.60	0.64 - 0.96	1.44 - 1.96
High P - Low K	0.40 - 0.60	0.64 - 0.96	0.36 - 0.54
Low P - High K	0.40 - 0.60	0.16 - 0.24	1.44 - 1.96
Low P - Low K	0.40 - 0.60	0.16 - 0.24	0.36 - 0.54

Table 8. Suggested Values for XLNT, XPLT, XKLT, HALFN, and PERIOD for Three Different Kinds of Soil.

Variable	Sandy Loam	Silt Loam	Clay Loam
XNLT	0.16 - 0.24	0.08 - 0.12	0.072 - 1.108
XPLT	0.28 - 0.42	0.20 - 0.30	0.16 - 0.24
XKLT	0.28 - 0.42	0.20 - 0.30	0.16 - 0.24
HALFN	212.0 - 318.0	252.0 - 378.0	292.0 - 438.0
PERIOD	10 - 14	12 - 18	14 - 22

Check/Inputs and Initialize Variables

This section of the subroutine GROW uses the input variables that define the status of a crop and soil nutrients (N, P, K) for a field at the beginning of the first day of a simulation. As each input variable is used, checks are performed; and, if an input variable is not within a prescribed range, it is corrected by the program which prints a message to that effect.

Program Section of Grow

The program section of subroutine GROW consists of nested DO loops. A day loop is nested within a run loop. The variables needed to fill the 32 arrays that are returned to the main program are updated daily in the day loop, with one exception. The exception is that the quantity of crop material on a field at the beginning of a run is determined and stored in an array before the program enters the day loop. Assimilation of the other variables into the arrays is accomplished at the end of the run loop (i.e., after completion of the day loop).

Day Loop

The day loop extends from the first through the last day of each consecutive run. Various subroutine functional relationships, primarily those listed in Table 5, are updated as independent variables and mathematically logical relationships are used to make the necessary determinations. When the last day of each successive run has been completed, subroutines are called to summarize the data for the just-completed run. These data are stored in the 33 data arrays. When all runs on all fields during all years have been completed, these data arrays are returned to the main program where the final output array will be created and written onto a data set. Four subroutines (SUM 1-4) summarize the data (Table 4) and store them in the proper arrays (Smith, et. al., 1984, 1985). The four subroutine summaries are the air temperature regime, rainfall and drought regimes, crop material harvested (grazed), nutrient utilization, and crop material remaining on a field at the end of each run.

ATTRIBUTES

The initial concern in the process of developing a dynamic simulation model is to discern the attributes whose changes need to be followed as time progresses.

Dry matter accumulation is one attribute used to signify the growth of crops. This attribute is especially appropos for studying the interaction of crop growth and grazing animals, because accumulated dry matter is also the attribute which is consumed by grazing animals. Grazing animals remove the leaves which are the organs that the crop uses to manufacture carbohydrates needed for growth. This interaction affects management decisions concerning the number of animals per unit of land area and the time intervals during the growing season of the crop when the animals are allowed to graze.

The crop growth portion of PASTURE is modified to allow the growth of a particular crop divided to encompass four partitions: (1) new growth, (2) old growth, (3) dead growth, and (4) stored carbohydrates utilized by the crop for regenerative growth. These partitions reflect the physiological development of a crop canopy and greatly improve the ability of the model to simulate photosynthetic activity and selective grazing.

New Growth Partition

The new growth partition is bounded by a minimum and a maximum physiological age of the growth. The minimum physiological age is zero physiological days old, and the maximum is the physiological age in days when the leaves of the crop are mature and begin senescence. This physiological period from minimum to maximum is analogous to the photosynthetic life of an individual leaf. The new growth partition is characterized by new growth being added each chronological day due to a daily growth rate. Even though this added growth is one chronological day old, it is not necessarily one physiological day old. The continuous flow of the mathematical-logic to simulate the physiological aging of the material in the new growth partition is predicated on the air temperature regime which is encountered on each successive chronological day.

When the physiological age in days of the material in the new growth partition exceeds the photosynthetic file in days of a typical individual leaf, then the material moves from the new growth partition to the old growth partition. Material may also be removed from the new growth partition by harvesting.

Old Growth Partition

The old growth partition is bounded by a minimum and a maximum physiological age of the growth. The minimum is the physiological age in days when the leaves of the crop are mature and begin senescence.

The maximum is the physiological age in days when senescence of the crop leaf is essentially complete. This physiological period for minimum to maximum is analogous to the time required for a mature leaf to die and turn from green to brown. Generally, this physiological period will be about the same as the photosynthetic life of the leaf; i.e., the maximal physiological age of the old growth will usually be about twice the maximum physiological age of the new growth.

Unlike the new growth partition, the old growth partition doesn't have material added to it on each chronological day. Material is added intermittently by being transferred from the new growth partition when its physiological age exceeds the photosynthetic life in days of a typical individual leaf; i.e. physiological aging. Material may be removed from the old growth partition by transfer to the dead growth partition or by harvest.

Dead Growth Partition

The dead growth partition is bounded by a minimum physiological age which is the physiological age in days when senescence of the crop leaves is essentially complete.

Material is added to the dead growth partition intermittently by being transferred from the old growth partition when its physiological age exceeds the age when senescence of the crop leaves is essentially complete. The material in the dead growth partition, unlike the new and old growth partition, doesn't contribute to crop growth, and the physiological aging is assumed to advance chronologically. Unlike the new and old growth partitions, the dead growth partition is not bounded by a maximum physiological age, consequently the material is not transferred to another partition. Material is removed from the dead growth partition due to weathering losses and by harvesting.

Non-Harvestable Stored Reserved Partition

In concept, the stored reserves partition is where the crop stores reserve carbohydrates for use in regenerative growth. This partition is visualized as being at the base of the crop. In so far as the mathematical-logic is concerned, the material in this partition is used only to support growth and is not available for harvest. The stored reserves partition, unlike the new, old, and dead growth partition, is not characterized by its physiological age. Rather, this partition is characterized only by the quantity of material in kg/ha or lb/ac which is in the partition at any time.

The mathematical-logic simulates regrowth as a function of stored reserves which are affected by frequency of defoliation, and affords the opportunity to simulate stand reductions due to winter and drought kill.

Daily Growth Rate

The growth rate of a crop is the time rate of change in the accumulated dry matter, usually expressed in units of kilograms of dry matter per hectare per day. These units are compatible with the photosynthetic process which occurs on a diurnal period.

Crop growth is a continuous function, and scientific investigations have verified that the function is a sigmoid (Thompson, 1942). The independent variable in this function is physiological time, not chronological time. Physiological time progresses only when photosynthesis is active, i.e., when the crop is growing.

Growth rate is the slope of the growth function and is also a continuous function. A rate cannot be observed, i.e., measured directly. The quantity of accumulated dry matter at a point in time can be measured. However, the rate at which this quantity is changing must be calculated. If the growth function is an algorithm, i.e., an accumulated growth by time matrix, growth rate is calculated as the difference between two successive elements of the algorithm divided by the time elapsed between these elements. This calculation gives an average growth rate and is a discontinuous or stepwise function. On the other hand, if the growth function is continuous (the case in this model), growth rate is the derivative of the growth function and is, itself, a continuous function (Smith and Loewer, 1981).

The daily growth rate is a function of the genetic potential of the crop, air temperature, day length, rainfall, leaf area, availability of nutrients (nitrogen, phosphorus, and potassium), and quantity of nonstructural carbohydrates in reserve.

Genetic Potential of the Crop

Each crop whose growth is to be simulated is described by different parameters which define the genetic potential of the crop (Table 2). These parameters are defined and have to be provided when a simulation is to be conducted with the model.

Air Temperature

Crops are seasonal, as exemplified by the accepted characterization of perennial grasses as cool-season and warm-season grasses (Heath et. al., 1985). The predominant environmental attribute which distinguishes the different seasons is air temperature. Crops are seasonal, because they are genetically adapted to grow in certain temperature regimes.

Plant physiologists (Salisbury and Ross, 1969) have documented three characteristic air temperatures as they relate to crop growth. There is an optimum air temperature where the growth rate of each crop is optimized which is bounded by minimum and maximum temperatures below and above at which growth does not occur. These three characteristic air temperatures and the optimum growth rate are four parameters which partially characterize the genetic potential of each crop whose growth is to be simulated. Growth is also a function of photosynthesis which occurs during a diurnal period which the model is capable of simulating. Diurnal air and soil temperature variation (Penrod, 1960 and Smith, et. al., 1968) is determined by four environmental parameters; (1) the minimum daily air temperature which occurs at sunup, (2) the maximum daily air temperature which occurs at solar noon, (3) the mean daily air temperature which occurs at sundown, and (4) the daylength (hours from sunup to sundown) which is calculated as a function of latitude. This accumulated growth for each day is the daily growth rate per day based on the air temperature regime.

Rainfall

A unique method was developed to reflect the effect of rainfall of the growth rate of crops. The method involves comparisons of actual daily rainfall and actual accumulated daily rainfall with normal daily rainfall and normal accumulated daily rainfall. Actual daily rainfall and normal monthly rainfall are input variables. The normal monthly rainfall values are used to calculate the normal daily rainfall. The actual daily rainfall and normal daily rainfall are used to obtain effective values of daily rainfall and accumulated daily rainfall to use in calculating a rainfall factor. The effective actual and normal quantites of rainfall are used to calculate ratios to reflect the current conditions and the antecedent conditions with respect to rainfall. The rainfall factor is calculated each day of simulation and is multiplied by the optimum growth rate to revise this parameter to reflect an altered growth regime.

Leaf Area

The daily growth rate as a function of air temperature and rainfall cannot be achieved unless there is enough leaf area to intercept sufficient solar energy to support photosynthesis. Functions simulate and relate the proportion of this potential daily growth rate which can be achieved to the quantity of dry matter in the new and old growth partitions (photosynthetically active partitions). These functions are based on two crop parameters: (1) the proportion of the potential daily growth rate which can be achieved when the crop first emerges from seed or other organs which initiate growth before photosynthesis is active; (2) the minimum quantity of accumulated dry matter which provides enough leaf area to support the full daily growth rate in an optimum temperature regime.

Potential and Actual Daily Growth Rates

A potential daily growth rate is determined as a function of air temperature, daylength, and rainfall, and then an actual daily growth rate is determined by multiplying the potential daily growth rate by the leaf area factor. The daily growth rate represents the assimilation of non-structural carbohydrates by the photosynthetic activity of the new and old growth crop partitions. The basic premise is that the part of the daily growth rate which is due to the new growth partition (growing leaves) is converted to structural carbohydrates, and that the part which is due to the old growth partition (mature leaves) is stored as non-structural carbohydrates. The part of the actual daily growth rate which is due to the new growth partition is available to be added to the new growth partition provided sufficient nutrients (N, P, K) are available to manufacture the structural carbohydrates. The stored non-structural carbohydrates might also be utilized to manufacture structural carbohydrates which can also be added to the new growth partition provided sufficient nutrients are available.

Availability of Nutrients

A growing crop must obtain nutrients from its environment (the soil and atmosphere) in order to utilize the non-structural carbohydrates which are synthesized during photosynthesis for the manufacture of structural carbohydrates (growth). The primary nutrients required by a growing crop are nitrogen, phosphorus, and potassium.

The mathematical-logic for determining the quantity of each of these nutrients required by a growing crop is based upon the premise that a crop will attempt to maintain the proper concentration of each of these nutrients in the cell content. With this premise as a basis, mathematical-logic is used to determine the required quantities of nitrogen, phosphorus, and potassium to manufacture the quantities of structural carbohydrates. These required quantities of nutrients are compared to the quantities which are available. If one or more of the required quantities is greater than that which is available, then the daily growth rates C and H are reduced to reflect the daily growth rate which can be produced with the most limiting nutrient. An accumulative accounting is maintained of the quantities of each nutrient used by a growing crop. The mathematical-logic also allows a crop to use more potassium than is required for growth, if it is available: i.e., potassium enrichment is a known phenomenon which is addressed in this model.

An accounting of the available nutrients is maintained, and each nutrient has a minimum quantity which is available at the beginning of each day. Nutrients are removed from the available supply by a growing crop in accordance with that which is required for growth. Nitrogen is also removed by leaching. Nutrients can be added to the available supply by scheduling fertilizer applications at specific times during a simulation.

Storing and Utilizing Non-Structural Carbohydrates

The fundamental concept used as the basis for developing this mathematical-logic is that photosynthesis uses solar energy to synthesize non-structural carbohydrates which are stored in a growing crop. Non-structural carbohydrates are stored in various parts of a growing plant: (1) new growth partition to simulate storage in growing leaves, (2) old growth partition to simulate storage in stems and mature leaves, (3) dead growth partition to simulate dead leaves and stems, and (4) stored reserves partition to simulate non-harvestable storage in roots and stolons. The logic maintains a record of the quantity of non-structural carbohydrates in each of these storage locations and adapts quantities each successive chronological day. Subsequently, these non-structural carbohydrates are utilized by the crop for manufacturing structural carbohydrates (growth). Maximum utilization of this resource occurs during the vegetative growth phase of crop development, and maximum storage occurs during the senescence phase. This resource may also be used to produce seed (grain) instead of vegetative growth. A marked delineation of the vegetative growth and the senescence phase does not occur within the plant. The mathematical-logic accommodates this simultaneous occurrence. This logic and the use of photosynthetically active parts of a growing crop in the growth equations significantly enhance the crop growth model as a tool for characterizing the nutritive value of a growing crop for grazing beef animals.

Nutritive Value for Beef Animals

The nutrition of growing crops contributes to what a grazing animal will consume and what parts of the crop will be consumed. The parts of the crop and the quantity of each part consumed affects the growth rate of the crop.

The nutritive value of a crop is directly related to the digestibility of the crop material. Digestibility is related to the cell wall content of the crop material and the digestibility of the cell wall material. The digestibility of the cell wall material is related to the physiological age of the material. The cell content is assumed to be completely digestible. The nitrogen content of the crop material is also an important attribute which is related to the nutritive value of a crop. These attributes are expressed as fractions of the quantity of material in each partition. Mathematical-logic has been developed and programmed into the crop growth model to simulate these dynamic changes.

Crop Harvest

The model contains mathematical-logic to simulate two different methods of harvesting. Daily harvesting scheduled to simulate grazing and event harvesting scheduled to simulate hay harvest.

Three options are available for specifying the quantity to be harvested each day during periods of daily harvesting. One option is to specify a certain quantity of dry matter to be harvested each day. This option is useful for evaluating a constant grazing rate. A second option is to specify a fraction of the available dry matter to be harvested each day. This option is used for evaluating a variable grazing rate. A third option is used to specify a fraction of the daily growth to be harvested each day. This option is especially useful for evaluating grazers who select only the new growth which occurs each day. A harvesting efficiency may be specified for each of these harvest options to reflect the losses which would be encumbered by certain grazers.

Daily harvesting is programmed to reflect selective grazing. Selectivity is based upon the digestibility of the dry matter in the three growth partitions. A fraction of the quantity to be harvested each day is harvested from each partition in the proportion to the fraction of the total available digestible dry matter which is contained in each partition.

Two options are available for specifying the quantity to be harvested during each scheduled event harvest. One option is to specify a certain quantity of dry matter to be harvested. This option is used for evaluating a system whereby only a predetermined quantity of forage is harvested and the balance is left for grazing. A second option is to specify a fraction of the available dry matter to be harvested. This option may be used for evaluating the quantities of harvested feed which would be expected with different harvesting methods. A harvesting efficiency can also be specified for each of these harvest options to reflect the losses which would result from different harvesting machines. Event harvesting removes dry matter from the three growth partitions in proportion to the fraction of the total available dry matter which is contained in each partition.

GRAZING MANAGEMENT SIMULATION

A simulated tall fescue grazing management study (Smith, et. al., 1985) was conducted at Lexington, KY utilizing two main management variables. One was the number of fields: one, two, three, four, and seven. Except for one field, which has no rest period, each field had a constant rest period of 30 days. The other main variable was simulated grazing rate; 44.0, 50.4, 56.0, and 61.6 kg/ha/da. Two grazing seasons during each year were used in the simulation; 15 April to 13 July and 3 October to 31 December. The main source of random variability was the climatological data for the years 1975-1979. Simulated output data (RUNS) were provided for each consecutive 10-day interval (9 RUNS for each grazing season). A split-plot experimental design was used as the statistical model for analyzing the simulated data.

Spring Grazing

These simulated data indicate that a one-field system with a grazing rate of 61.6 kg/ha/da is equal to or superior to other combinations of numbers of fields and grazing rates (Table 9). This system resulted in a simulated mean harvest rate of 53.1 kg/ha/da and a potential tissue production for the 90-day period of

Table 9. Summary of Simulated Results for the Spring Grazing Season on Fescue with Respect to the Main Variables for Different Attributes.

	Number of Fields				
Attribute	1	2	3	4	7
Harvest Rate (kg/ha/da)	53.08*[a]	50.80[c]	52.53[a,b]	51.29[b,c]	52.12[a,b]
Digestible	65.96[b]	66.25[a]	65.60[c]	65.70[c]	65.62[c]
Dry Matter (kg/ha/da)	35.03[a]	33.61[c]	34.46[a,b]	33.90[b,c]	34.22[b,c]
Cell Content	1.85[b]	1.88[a]	1.87[a,b]	1.88[a]	1.87[a,b]
Nitrogen (kg/ha/da)	0.99[a]	0.95[b]	0.99[a]	0.96[a,b]	0.97[a]
Potential Tissue Production (kg(ha/da)	2.46[a,b]	2.37[c]	2.48[a]	2.42[b]	2.45[a,b]

	Simulated Grazing Rate (kg/ha/da)			
Attribute	44.0	50.4	56.0	61.6
Harvest Rate (kg/ha/da)	44.70[d]	50.16[c]	54.94[d]	58.30[a]
Digestible	65.90[a]	65.89[a]	65.82[a]	65.70[a]
Dry Matter (kg/ha/da)	29.46[d]	29.52[c]	36.16[b]	38.34[a]
Cell Content	1.81[d]	1.85[c]	1.89[b]	1.94[a]
Nitrogen (kg/ha/da)	0.81[d]	0.93[c]	1.04[b]	1.12[a]
Potential Tissue Production (kg(ha/da)	2.05[d]	2.33[c]	2.59[b]	2.78[a]

*Means, for each variable and attribute, with the same letter (superscript) are not significantly different (Waller-Duncan K-Ratio t Test).

approximately 250 kg/ha/da. The model logic used in the simulation of potential tissue production is based on ruminant protein requirements (ARC, 1980) and digestible dry matter and dietary nitrogen (Greenhalgh, 1981). An analysis by RUNS for various attributes indicates a gradual and expected decline for cell content nitrogen and potential tissue production except during periods of mid to late May (Table 10).

Fall Grazing

Data from this simulation indicates that a three-field system and a grazing rate of 61.1 kg/ha/da is equal to or superior to other main variable combinations (Table 11). An analysis of means for variables and attributes imply that there was a harvest rate of 56.5 kg/ha/da and a mean potential tissue production of 2.46 kg/ha/da. The simulated potential tissue production rate was about 22.4 kg/ha lower than the spring grazing period during the 90-day period. Potential tissue production decreased throughout the grazing season (Table 12).

CONCLUSION

The need to evaluate responses in the agricultural sciences from a systems perspective has become increasingly apparent. Model simulation, or components of models which may be run separately, make it possible to evaluate and understand fundamental relationships that may not be evaluated without model simulation. For example, simulation may delineate physiological and chronological relationships of plant and animal growth, combined effects of environment and maturity on forage yield, digestibility, and utilization. Simulation may also be used in designing research prior to the initiation of field studies. Various forage-beef production systems may result in different seasonal feed supply distributions contributing to varying beef production which may lead to increased and more efficient resource substitution. Mathematical simulation can never replace individual experienced judgement nor can it elucidate cause-effect relationships that cannot be quantified, however, predictive and descriptive models may provide information to assist in understanding past events, analyzing results and alternatives, and anticipating future events, and transforming this information into a more understandable form and time saving effort (Barnes, et. al., 1985).

Table 10. Summary of Simulated Results for the Spring Grazing Season on Fescue with Respect to the Run Variable for Different Attributes.

RUN	Harvest Rate (kg/ha/da)	Attribute				
		Digestible Dry Matter		Cell Content Nitrogen		Potential Tissue Production (kg/ha/da)
		(%)	(kg/ha/da)	(%)	(kg/ha/da)	
April 15-24	51.79*[b]	64.32[d]	33.33[c]	2.08[a]	1.09[a]	2.58[a]
April 25 - May 4	52.89[a,b]	65.02[c,d]	34.41[b]	1.99[b]	1.05[a]	2.52[a]
May 5-14	50.46[c]	64.91[c,d]	32.85[c]	1.97[b]	0.99[b]	2.39[b]
May 15-24	52.88[a,b]	64.79[d]	34.27[b]	1.89[c]	1.00[b]	2.50[a]
May 25 - June 3	52.19[a,b]	65.63[b,c]	34.26[b]	1.88[c]	0.99[b]	2.50[a]
June 4-13	52.64[a,b]	67.45[a]	35.47[a]	1.72[e]	0.91[c]	2.37[b,c]
June 14-23	53.17[a]	67.62[a]	35.94[a]	1.69[e]	0.90[c]	2.33[b,c]
June 24 - July 3	52.17[a,b]	66.32[b]	34.60[b]	1.79[d]	0.93[c]	2.40[b]
July 4-13	49.90[c]	66.32[b]	33.14[c]	1.81[d]	0.90[c]	2.30[c]

*Means, for each attribute, with the same letter (superscript) are not significantly different (Waller-Duncan K-Ratio t Test).

Table 11. Summary of Simulated Results for the Fall Grazing Season on Fescue with Respect to the Main Variables for Different Attributes.

	Number of Fields				
Attribute	1	2	3	4	7
Harvest Rate (kg/ha/da)	52.33*[a]	49.00[c]	52.85[a]	51.14[b]	49.13[c]
Digestible	66.38[a]	65.22[b]	65.99[a,b]	66.03[a]	66.10[a]
Dry Matter (kg/ha/da)	34.78[a]	32.29[c]	34.89[a]	33.80[b]	32.55[c]
Cell Content	1.66[a]	1.67[a]	1.68[a]	1.68[a]	1.67[a]
Nitrogen (kg/ha/da)	0.86[a,b]	0.83[c]	0.88[a]	0.86[b]	0.82[c]
Potential Tissue Production (kg(ha/da)	2.26[b]	2.16[c]	2.32[a]	2.25[b]	2.15[c]

	Simulated Grazing Rate (kg/ha/da)			
Attribute	44.8	50.4	56.0	61.6
Harvest Rate (kg/ha/da)	44.14[d]	49.34[c]	53.51[b]	56.56[a]
Digestible	65.90[a]	65.94[a]	65.76[a]	66.17[a]
Dry Matter (kg/ha/da)	29.10[d]	32.55[c]	35.37[b]	37.63[a]
Cell Content	1.68[a]	1.68[a]	1.67[a]	1.65[a]
Nitrogen (kg/ha/da)	0.74[d]	0.83[c]	0.90[b]	0.94[a]
Potential Tissue Production (kg(ha/da)	1.94[d]	2.16[c]	2.35[b]	2.46[a]

*Means, for each variable and attribute, with the same letter (superscript) are not significantly different (Waller-Duncan K-Ratio t Test).

Table 12. Summary of Simulated Results for the Fall Grazing Season on Fescue with Respect to the Run Variable for Different Attributes.

RUN	Harvest Rate (kg/ha/da)	Attribute				
		Digestible Dry Matter		Cell Content Nitrogen		Potential Tissue Production (kg/ha/da)
		(%)	(kg/ha/da)	(%)	(kg/ha/da)	
October 3-12	53.19*[a]	63.64[f]	33.85[b]	2.10[a]	1.11[a]	2.67[a]
October 13-22	53.19[a]	64.65[e]	34.40[b]	1.95[b]	1.04[b]	2.60[b]
Oct. 23 - Nov. 1	53.19[a]	66.39[c,d]	35.34[a]	1.77[c]	0.94[c]	2.45[c]
November 2-11	53.19[a]	67.59[a]	35.97[a]	1.62[d]	0.86[d]	2.32[d]
November 12-21	52.85[a]	68.12[a]	36.01[a]	1.53[e,f,g]	0.81[e]	2.16[e]
Nov. 22 - Dec. 1	53.17[a]	67.41[a,b]	35.86[a]	1.51[f,g]	0.80[e]	2.15[e]
December 2-11	52.83[a]	66.73[b,c]	35.27[a]	1.50[g]	0.78[e,f]	2.13[e]
December 12-21	50.04[b]	65.57[d]	32.83[c]	1.53[e,f]	0.76[f]	2.06[f]
December 22-31	36.32[c]	63.38[f]	23.46[d]	1.54[e]	0.56[g]	1.52[g]

*Means, for each attribute, with the same letter (superscript) are not significantly different (Waller-Duncan K-Ratio t Test).

REFERENCES

ARC. 1980. The Nutrient Requirements of Ruminant
Livestock. Commonwealth Agricultural Bureaux,
Farnham Royal. Slough 2L2 3BN, ENGLAND.

Barnes, R.F., C.O. Little, and M.A. Brown. 1985.
"Influence of Systems Modeling on Agricultural
Policy." In Simulation of Forage and Beef Produc-
tion in the Southern Region. V.H. Watson and C.M.
Wells, Jr. (Ed.), pp. 133-135, Southern Cooperative
Series, Bulletin 308.

Binger, R.L., L.G. Wells, and T.C. Bridges. 1980. "A
Simulation Model of Resource Management in the
Production of Burley Tobacco." ASAE Paper No. 80-
3553, ASAE, St. Joseph, MI 49085.

Brown, B. 1982. "Computer Simulation of Plant and
Animal Growth." Nebraska Beef Cattle Report, University of Nebraska, MP 43:5.

Brown, R.H., and R.E. Blazer. 1968. "Leaf Area Index
in Pasture Growth." Herbage Abstracts. 38:1-9.

Burton, G.W., W.S. Wilkerson, and R.L. Carter. 1969.
"Effect of Nitrogen, Phosphorus, and Potassium Levels and Clipping Frequently on the Forage Yields
and Protein, Carotene, and Xanthophyl Content of
Coastal Bermudagrass." Agronomy Journal. 61:60-
63.

Davidson, J.L., and F.L. Milthorpe. 1964. "Carbohyd-
rate Reserves in the Regrowth of Cocksfoot (Dactyl-
is glomerata L.)." Journal of British Grassland
Society. 20:15-18.

Ewen, L.S. 1980. "Growth of Fescue and Red Clover as
Influenced by Environment and Interspecific Compe-
tition." M.S.A.E. Thesis, University of Kentucky,
Lexington.

Greenhalgh, J.F.D. 1981. "A Forward Look at Technical
Possibilities for Grassland. Grassland in the
British Economy." CAS Paper 10. Reading: Center
for Agricultural Strategy, pp. 51-63.

Heath, M.E., D.S. Metcalfe, and R.F. Barnes. 1985.
Forages. 4th Edition, The Iowa State University
Press, Ames, Iowa.

136

Murata, U.J., J. Iyama, and T. Honma. 1965. "Studies on the Photosynthesis of Forage Crops (IV)." Proceedings of the Crop Science of Japan. 34:154-155.

Neels, D.P. 1981. "Simulation of Alfalfa Growth and Harvest for Improved Machinery Management." M.S. Thesis, University of Nebraska, Lincoln.

Penrod, E.B. 1960. "Variation of Soil Temperature at Lexington, KY." Engineering Experiment Station Bulletin, No. 57, University of Kentucky, Lexington, KY.

Salisbury, F.B., and C. Ross. 1969. Plant Physiology. Wadsworth Publishing Company, Inc. Belmont, CA.

SAS. 1979. SAS Users' Guide, 1979 edition. SAS Institute, Inc., Box 8000, Cary, NC.

Smith, E.M., T.H. Taylor, and L. Brown. 1968. "Interpretation of Diurnal Variation in Soil Temperatures." Transactions of ASAE. 11:195-197.

Smith, E.M., and O.J. Loewer. 1981. "A Nonspecific Crop Growth Model." ASAE Paper No. 81-4013, ASAE, St. Joseph, MI 49085.

Smith, E.M., and O.J. Loewer. 1983. "Mathematical-logic to Simulate the Growth of Two Perennial Grasses." Transactions of ASAE. 26:878-883.

Smith, E.M., L.M. Tharel, M.A. Brown, C.T. Dougherty, and K. Limbach. 1984. "A Simulation Model for Managing Perennial Grass Pastures I. Model Structure." Agricultural Systems. 17:155-180.

Smith, E.M., L.M. Tharel, M.A. Brown, G.W. Burth, C.T. Dougherty, S.L. Fales, V.H. Watson, and G.A. Pederson. 1985. "The Plant-Growth Component." In Simulation of Forage and Beef Production in the Southern Region. V.H. Watson and C.M. Wells, Jr. (Ed.), pp. 9-36, Southern Cooperative Series, Bulletin 308.

Spedding, C.R.W. 1975. "Grazing Systems." Proceedings III World Conference on Animal Production. R.L. Rein (Ed.), Sydney University Press, pp. 145-157.

Thompson, D.W. 1942. Growth and Form. MacMillan Company, New York, NY.

VI

Use of the Kentucky Beef-Forage Model in Economic Analysis

Lucas D. Parsch, Otto J. Loewer,
and David H. Laughlin

INTRODUCTION

In recent years, researchers have shown interest in evaluating the risk-return trade-offs of alternative agricultural production management strategies. One of the key ingredients which enables researchers to conduct such analyses is the probability distribution of the output variable by which system performance is measured.

For most crop-based agricultural systems, weather has been hypothesized to be the major source of production risk--defined here as performance variability.[1] If this hypothesis is correct, one of the major benefits resulting from the use of biological-phenological models in risk research is that they are capable of simulating experimental data from which the probability distribution of the performance measure can be estimated. Often, this information is not available from field experiments.

The purpose of this paper is to demonstrate use of the biological-phenological Kentucky Beef Forage Model (GRAZE) in generating output which can be used in economic risk-return analysis. Under the assumption that weather is the major source of production risk for a summer pasture steer grazing system, the performance of

Department of Agricultural Economics, University of Arkansas; Department of Agricultural Engineering, University of Arkansas; Department of Agricultural Economics, Mississippi State University.

[1]Another major category of risk in agriculture is price or market risk, which is not addressed in this paper.

30 alternative production management strategies is evaluated under 10 alternative "states of nature" or weather scenarios. Comparison of the outcome distributions indicates that those strategies resulting in the highest expected net weight gain per hectare do not result in the highest expected net returns per hectare.

SIMULATION OF GRAZING SYSTEM PERFORMANCE AS A FUNCTION OF MANAGEMENT AND ENVIRONMENTAL FACTORS

The 30 management strategies evaluated in this study reflect a broad spectrum of combinations of both grazing intensity and rotational grazing. Each of the 30 strategies is defined within the GRAZE model by setting the level of three input variables:

(a) the stocking rate, i.e. the number of steers grazed per hectare of pasture;
(b) the number of parcels or grazing fields that the total pastureland is divided into; and,
(c) the rotation period, or length of time that each of the parcels is grazed before the animals are moved to the next parcel.

The GRAZE model consists of a phenological plant growth-composition model (Smith, et al., 1985); a physiological animal growth-feed intake model S-156 (Loewer, et al., 1985a); and, a plant-animal interface model which describes the logic of selective grazing as a function of environment (Loewer, et al., 1985b). For any given management strategy, the GRAZE model simulates net animal growth and performance (net weight/ha) as a function of the forage quantity and quality available to the animal. For each strategy, GRAZE model output (net animal weight produced per hectare) is then evaluated for its economic performance (net returns per hectare) by calculating the specified costs and revenues associated with each strategy.

Simulation of each of the 30 management strategies over ten "states of nature" or weather scenarios is accomplished by using ten years of daily historical weather data as "driver" variables for both the crop and animal submodels. Hence, final model output consists of 10 alternative measures of animal performance (net weight gain/ha) and 10 corresponding measures of economic performance (net returns/hectare) for each management strategy analyzed.

Under the assumption that the historical weather data is representative of the long-run range of temperatures and precipitation experienced at a site, these 10 performance measures can be viewed as ten points on the sample frequency distribution for each strategy. Subsequent procedures to convert this information into an appropriate form for risk analysis can then include calculation of sample moments, curve-fitting of frequency distributions (Anderson, et al., 1977), estimation of fractiles (Schlaifer, 1959), or, in the development of empirical cumulative distribution functions (Hogg and Tanis, 1977).

Use of the GRAZE model in this fashion permits the economist to design experiments which provide the following information necessary in risk research:

(a) a measurement of grazing system response and economic performance as a function of alternative management strategies (system controllable input); and,

(b) an assessment of the sensitivity of each management strategy to weather risk (system uncontrollable inputs and their likelihood of occurrence).

EXPERIMENTAL DESIGN

Scenario Description

The simulation experiment was designed to evaluate summer grazing of steer calves on grass pasture at Booneville, Arkansas. All scenarios began with 213 kg. steer calves placed onto common bermuda grass (cynodon dactylon) pasture on June 1 of each year. High levels of nitrogen (112 kg/ha) were applied to the pasture on April 1 and June 15 of each simulated year; levels of P_2O_5 and K_2O were set so as to be non-limiting.

All steers were backgrounded on the pasture for a maximum of four months and were provided with no supplemental feed except for minerals and water. On October 1 of each year, the simulation was terminated and the animals were sold. Only under two conditions were animals sold prior to October 1:

(a) whenever a negative weight gain was observed for four consecutive weeks; or,

(b) whenever total animal weight loss was greater than 15% of the initial weight of the animals.[2]

_____

[2]A 15% weight loss reflects severe deficiencies and a reduced potential for economic recovery. This weight loss initial value was suggested by Dr. C.J. Brown (1984).

These arbitrary feed-back controls reflect a rational manager's deviation from the given October 1 selling date in order to reduce potential losses whenever environmental factors result in a decline in animal performance due to the quantity of available pasture.

Management Strategies

Thirty alternative management strategies reflecting grazing intensity and rotational grazing were identified based on an experimental design described by Smith, et al. (1985). A single management strategy for this study consisted of any combination of three management factors. The three management factors, and the alternative levels at which each was evaluated are described below:

(a) stocking rate (SRATE) = number of animals/hectare of pasture (2, 7, or 12 head/ha).
(b) number of fields (NFIELDS) = number of grazing parcels that the total pasture area is divided into (1, 2, 3, or 4 parcels).
(c) rotation/rest period (ROTPD) = number of days each parcel is grazed before the animals are moved to the next parcel (continuous, 7, 14, or 21 days).

Whereas the three stocking rates reflect a spectrum of alternative grazing intensities, the rotational grazing aspect of this study is defined by any combination of the number of fields and the length of rotation period. Table 1 demonstrates that the four levels of NFIELDS and ROTPD give rise to ten alternative rotation patterns. Since any rotation pattern can be combined with one of three stocking rates, a total of 30 management strategies can be identified.

Weather Data

Ten years of daily historical weather data (maximum and minimum temperature, precipitation) for Booneville, Arkansas for the years 1972-1981 inclusive were used to simulate ten alternative "states of nature" for each of the 30 management strategies defined above. Rainfall distribution patterns for the four-month summer season are shown in Figure 1. The 10-year period reflects 4-month rainfall totals ranging between 24.01 inches (1973) and 8.62 inches (1976). Rainfall distribution over the months of July and August for the same period ranged between 11.52 inches (1981) and 0.6 inches (1980).

Table 1. Number of sequential days a pasture is not
          grazed for an alternative number of fields
          and grazing periods.

Number of fields:		1	2	3	4
Rotation Period:	C	0	-	-	-
	7	-	7	14	21
	14	-	14	28	42
	21	-	21	42	63

Note: Rotation period is the number of days grazed. C
represents continuous grazing. The blank values in row
1 and column 1 of Table 1 indicate that anything other
than continuous grazing is logically inconsistent for a
one-field system. By contrast, two, three of four-
field systems can be combined with any one of three
rotation periods each.

THE ECONOMIC MODEL

GRAZE model output for animal ending-weight was
used to calculate net returns (dollars/hectare) for
each of the 300 strategy years simulated. Net returns
were defined as the gross returns from the sales of the
animals minus selected costs associated with each
strategy, i.e.,

Net Returns = Gross Returns - Selected Costs

Because "selected costs" included only those costs
which might be affected by stocking rate, number of
fields, or rotation period, the net returns variable
measures the dollar contribution to overhead labor,
management, land, and overhead capital generated by
each strategy.

Gross Returns

Gross returns is the product of simulated animal
ending weight and the market price received, i.e.,

GR = [PBEEF(weight class, month)] * WT/HD * SRATE

where:

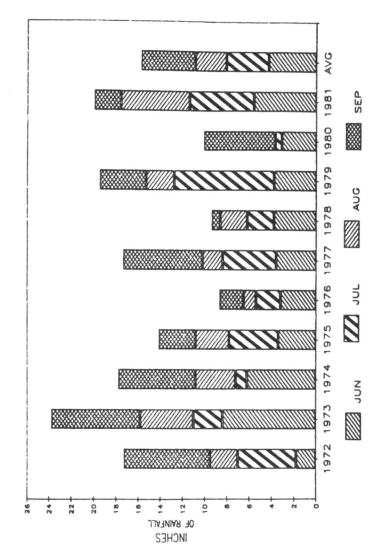

Figure 1. Rainfall distribution patterns for summer season, Booneville, Ark., 1972-1981

GR = gross returns, $/hectare
PBEEF = price received for beef, a function of weight
class and month of sale, $/kg
WT/HD = simulated animal weight, kg/head
SRATE = stocking rate

The market price received for the animals is the Arkansas auction (USDA-AMS) 10-year average price for No. 1 medium-frame steers indexed to 1983 price levels. Table 2 shows the 10-year June-October average price for four weight classes (300-700 pounds). For each animal ending weight, these prices were multiplied by the appropriate monthly price indexes to reflect simulated sales occurring at times other than October. Animals dropping below the 15% weight loss restriction received a 30% discount off the market price for the relevant month of sale.

Selected Costs

Cost categories computed for each simulated scenario are summarized in Table 3. The steer calf purchase price is the 10-year average Arkansas May auction price for No. 1 medium-frame steers indexed to the 1983 price level. Cost categories 2 through 5 reflect representative production and marketing costs for Arkansas cattle producers (Spaulding, et al., 1983). The interest charge (cost category 8) reflects the short-term cost of capital to carry the animals until they are sold.

Because fencing requirements increase as the number of grazing parcels (NFIELDS) is augmented, an annual charge for fencing is included. An annual estimated charge of $95 per quarter mile of fencing was converted to annual charges of $11.73, $14.66, $16.61 and $17.60 per hectare for 1, 2, 3 and 4-field systems respectively. Calculations per hectare were based on the fencing requirements for a total pasture area of 160 acres. Annual charges reflect the cost of materials, installation, and repair of a 5-wire steel-post system.

Labor costs in the model reflect the fact that labor requirements increase as the rotation period (ROTPD) decreases. For example, animals rotated every 7 days require twice the herding labor of grazing systems in which animals are rotated every 14 days. An arbitrary charge for 2, 4 and 6 herding-hours of labor at $5.00/hour was assessed for stocking rates (SRATE) of 2, 7 and 12 head. These herding charges reflect labor costs for rotating the animals from one field to the next for a 160-acre total pasture area. These charges convert to $0.15, $0.31 and $0.46 per rotation per hectare for the low, medium and high stocking rates, respectively.

Table 2. Monthly price index, and average price received for #1 steers, medium frame, 1983 dollars.[a]

| Weight Class | | Jun | Jul | Aug | Sep | Oct | Avg: 1973–1983 | |
lbs.	kg.						$/cwt	$/kg
300–400	136–181	1.01	1.00	1.02	1.00	.97	71.89	1.585
400–500	181–227	1.01	1.00	1.02	.99	.98	67.01	1.478
500–600	227–272	1.00	1.00	1.02	1.00	.98	63.65	1.403
600–700	272–317	1.00	1.01	1.01	1.00	.98	61.73	1.361

[a]Based on Arkansas auctions, 1973–1983. Average price is averaged over June–October and indexed to 1983 price levels. Animals dropping below 400 lbs. receive a 30% discount off the market price for the relevant month of sale.

Table 3. Estimated production and marketing costs for
          backgrounded calves, Arkansas, 1983 dollars.

Cost Category	Value
1. Purchase steer calf	213 kg. (470 lbs.) @ $1.553/kg. ($70.43/cwt.)
2. Selling costs	3% of gross returns
3. Buying costs	$2/head
4. Hauling cost (two ways)	$4/head
5. Animal health costs	$7/head
6. Annual fencing charge	Function of number of grazing parcels (NFIELD)
7. Labor (herding) cost	Function of stocking rate (SRATE), length of rotation period (ROTPD), and week of sale
8. Interest on operating capital	13%/year times number of months animal is carried

RESULTS

Simulated Animal Performance

Tables 4 and 5 provide selected sample statistics
for simulated net animal weight gain by management
factor. Table 4 indicates that net weight gain per
hectare increased with the stocking rate, but that it
was inversely related to the number of fields and the
length of the rotation period. Likewise, of the three
management factors, stocking rate exerted the greatest
influence on animal gain per hectare as is evidenced by
the large range of mean weight gains across stocking
rate levels in comparison to the corresponding values
for number of fields and rotation period.

Table 5 indicates that when the effect of stocking
rate is removed, the number of fields and length of
rotation period had little impact on weight gain at the
low and medium stocking rates. The small absolute
difference in mean weight gains at the low stocking
rate signifies that the steers could selectively graze
due to virtually no competition among them for available pasture.

Table 4. Sample statistics for simulated net animal
weight produced by stocking rate, number of
fields, and rotation period, kg/ha.

		$\bar{X}^a$	Min[b]	Max[c]	CV[d]
Stocking Rate	2	160	137	179	6
(hd/ha):	7	495	-236	632	33
	12	568	-622	1068	92
Number of	1	488	-103	1066	69
Fields:	2	462	-378	1036	71
	3	394	-523	1068	95
	4	340	-622	1057	112
Rotation Period	C	488	-103	1066	69
(days):	7	459	-307	1030	74
	14	401	-622	1057	92
	21	336	-604	1069	113

[a]Sample mean.

[b]Smallest simulated value.

[c]Largest simulated value.

[d]CV is the coefficient of variation expressed as a
percentage.

Table 5. Mean simulated net animal weight produced for
alternative stocking rates, number of fields,
and rotation periods.

		Stocking Rate					
		2/HA		7/HA		12/HA	
		KG/HA	KG/HD	KG/HA	KG/HD	KG/HA	KG/HD
Number of	1	160	80	531	76	773	64
Fields:	2	158	79	520	74	708	59
	3	160	80	494	71	530	44
	4	161	80	460	66	397	33
Rotation	C	160	80	531	76	773	64
Period	7	158	79	430	61	690	58
(days):	14	160	80	498	71	545	45
	21	161	80	446	64	401	33

By contrast, at the high stocking rate, weight gain per hectare significantly decreased as the number of fields or length of rotation period increased. A one-field continuous grazing system was least likely to force animals to compete with each other for available forage during summer drought periods when pasture growth was retarded. Severe competition under intensive grazing (12 head/ha) is indicated in the results shown in Table 5 for both 4-field and 21-day rotation periods. Net animal weight gain per hectare for these stocking rates is less than for the medium stocking rate (7 head/ha) with a similar number of fields or rotation period.

Simulated Economic Performance

Tables 6 and 7 present the economic results corresponding to the animal performance measures reported in Tables 4 and 5. The most noteworthy difference between the two sets of tables is that unlike net animal weight, net returns per hectare do not monotonically increase with the stocking rate. Although the high stocking rate (12 head/ha) made most efficient use of the pasture in that it resulted in the greatest mean animal weight gain per hectare, it resulted in a lower weight gain per animal. than did either the low (2 head/ha) or medium (7 head/ha) stocking rates (Table 5). Hence, at 12 head/ha, competition for available forage brought about reduced animal efficiency and lower net returns as well.

The decrease in net returns when stocking rate was increased provides a classic example of how excessive use of inputs (i.e. number of steers) reduces marginal value productivity to the point where it is less than marginal factor cost.[3] This point can be clarified in Table 8 which shows output for a single simulated year for three arbitrarily selected management strategies which differ only in their stocking rate levels. As stocking rate is increased from 7 to 12 head, marginal factor cost is approximately constant. Cost categories 8 through 13 are solely a function of stocking rate and comprise a large proportion of total production costs per hectare. By contrast, the reduced performance of each animal at the high stocking rate results in

---

[3]For a concise definition of marginal value product and marginal factor cost and their impact on profit-maximizing input levels, see Beattie and Taylor (1985, pp. 76-92).

Table 6. Sample statistics for simulated net animal
          returns by stocking rate, number of fields,
          and rotation period, $/ha.

		$\bar{X}^a$	s^b	Min^c	Max^d
Stocking Rate	2	17	13	-14	41
(hd/ha):	7	30	197	-1087	195
	12	-275	690	-2217	333
Number of	1	44	200	-679	333
Fields:	2	5	242	-1225	288
	3	-109	512	-2130	328
	4	-164	536	-2218	311
Rotation Period	C	44	200	-679	333
(days):	7	3	244	-988	275
	14	-87	470	-2130	311
	21	-184	569	-2218	328

[a]Sample mean.

[b]Sample standard deviation.

[c]Smallest simulated value.

[d]Largest simulated value.

Table 7. Mean simulated net returns for alternative
          number of fields and rotation periods by
          stocking rate, $/ha.

		Stocking Rate		
		2/HA	7/HA	12/HA
Number of	1	22	78	30
Fields:	2	16	61	-62
	3	16	31	-374
	4	17	-18	-491
Rotation	C	22	78	30
Period	7	13	67	-70
(days):	14	17	39	-317
	21	19	-31	-439

Table 8. Net returns above specified costs for 3
selected strategies for simulated year 1976.

| ITEM | (1 Field, Continuous Grazing, 3 Alternative Stocking Rates) Strategy | | |
	(1, C, 2)	(1, C, 7)	(1, C, 12)
1. Hd/Ha	2	7	12
2. Sale Date	Oct 1	Oct 1	Oct 1
3. Wt/Hd, Kg	292	293	271
4. Sale Price, $/Kg	1.334	1.334	1.375
5. Gross Wt/Ha, Kg	583	2054	3249
6. Net Wt/Ha, Kg	157	560	689
RETURNS			
7. Gross Returns, $/Ha (4x5)	$778.18	$2739.49	$4467.38
COSTS, $/Ha			
8. Purchase Steer Calf, $331.31/Hd	$662.62	$2319.17	$3975.72
9. Interest (5 Mo @ 13%)	35.89	125.62	215.35
10. Selling Cost (3%*7)	23.35	82.18	134.02
11. Buying Cost ($2/Hd)	4.00	14.00	24.00
12. Hauling Cost, 2-ways ($4/Hd)	8.00	28.00	48.00
13. Vet Costs ($7/Hd)	14.00	49.00	84.00
14. Annual Fencing Cost ($/Ha)	11.84	11.84	11.84
15. Labor Herding Cost	0	0	0
16. Total Selected Cost	$759.70	$2629.81	$4492.93
17. Net returns (7 minus 16)	$18.48	$109.68	$-25.55

greater increments to per hectare costs than to per hectare returns as stocking rate is increased from 7 to 12.

The above comparison between 7 and 12 head stocking rates does not imply that net returns per hectare exhibit a positive relationship with weight gain per head. Sample statistics shown in Table 9 for the same three management strategies demonstrate that higher mean weight gain per animal for the low stocking rate (2 head/ha) results in lower per hectare mean net returns than for the medium stocking rate (7 head/ha). Hence, obtaining the highest expected net returns per hectare reflects a trade-off between choosing those strategies which are animal efficient but which under-utilize the pasture (low stocking rate), as compared to those strategies which are pasture-efficient but which under-utilize the grazing animals (high stocking rate).

Table 9. Summary statistics for 3 selected strategies.[a]

| (1 Field, Continuous Grazing, 3 Stocking Rates) | | | |
Strategy	(1, C, 2)	(1, C, 7)	(1, C, 12)
Mean Net Returns, $/Ha	22	78	30
Standard Deviation	13	148	324
Minimum	-5	-333	-679
Maximum	41	177	333
Net Weight/Ha, (KG)	160	531	773
Gross Weight/Hd, (KG)	293	289	278
Net Weight/Hd, (KG)	80	76	64
Average Daily Gain, (KG)	.64	.61	.52

[a]Based on 10-years of simulated data.

Table 10 exhibits the top five performing strategies when ranked on the basis of expected net returns per hectare. It is noteworthy that these were not the same strategies which resulted in the highest mean animal weight gain per hectare. Both the high (12 head/ha) and low (2 head/ha) stocking rates which generated the greatest mean net returns per hectare ($29.75 and $22.41, respectively) were one field, continuous grazing systems.

Table 10. Top five strategies ranked by mean net
returns ($/Ha)

Rank	Strategy (Fields, Rot'n Days, Hd/Ha)	Mean $/Ha
1	1, Continuous, 7 Hd/Ha	$78.43
2	2, 7 Days, 7 Hd/Ha	74.91
3	2, 14 Days, 7 Hd/Ha	74.21
4	4, 7 Days, & Hd/Ha	62.49
5	3, 7 Days, 7 Hd/Ha	62.35

Note: When ordered by net animal weight/ha produced
(mean), these five strategies ranked 9, 7, 8, 10, and
11 respectively.

ASSESSMENT OF PROBABILITY CHARACTERISTICS
FOR RISK ANALYSIS

The impact of weather variability on simulated
animal and economic performance is demonstrated in
Tables 11 and 12. Table 11 shows the impact of weather
on animal production and ranks the "states of nature"
(historical weather data) by mean animal weight gain
across all management strategies. Poor distribution of
rainfall during some years (1972, 1980) and low total
rainfall amounts in other years (1976, 1978) are evidenced by a reduction in animal weight gains due to
reduced pasture growth. The broad range of reported
mean values for animal weight gain and net returns in
Tables 11 and 12 indicates that the influence of
weather on grazing system performance is significant.

Tables 11 and 12 provide no indication of how the
performance of individual management strategies is
affected by weather extremes. However, the coefficients of variation do suggest a broad range of
responses across the 30 management strategies for any
given weather scenario.

The differential impact of weather on the performance of three selected grazing management strategies is
demonstrated in Figure 2. The high stocking rate (12
head/ha) is shown to be the high risk strategy in that
its performance is highly variable as a function of
weather. By contrast, the low stocking rate (2 head/
ha) appears to be little affected by weather, and
hence, is relatively riskless. It is noteworthy that
the ranking of the three strategies changes from one
weather scenario to the next. In "good" years (1974,

Table 11. Sample statistics for simulated net animal weight produced by year, kg/ha.

		$\bar{X}$[a]	Min[b]	Max[c]	CV[d]
Year	1979	579	169	1068	59
	75	573	163	1023	59
	73	566	152	1057	60
	74	528	166	1066	59
	81	518	150	952	58
	77	509	156	968	59
	78	411	-32	803	58
	76	210	-532	689	141
	72	157	-622	586	213
	80	7	-604	269	3476

[a]Sample mean.

[b]Smallest simulated value.

[c]Largest simulated value.

[d]CV is the coefficient of variation expressed as a percentage.

Table 12. Sample statistics for simulated net returns by year, $/ha.

		Net Returns, $/ha			
		$\bar{X}$[a]	s[b]	Min[c]	Max[d]
Year	1979	150	109	-49	328
	75	119	93	-72	278
	73	109	87	7	311
	74	73	175	-678	333
	81	52	88	-205	186
	77	49	134	-450	206
	78	-57	222	-816	117
	76	-325	604	-1988	110
	72	-412	695	-2086	144
	80	-518	646	-2217	-5

[a]Sample mean.

[b]Sample standard deviation.

[c]Smallest simulated value.

[d]Largest simulated value.

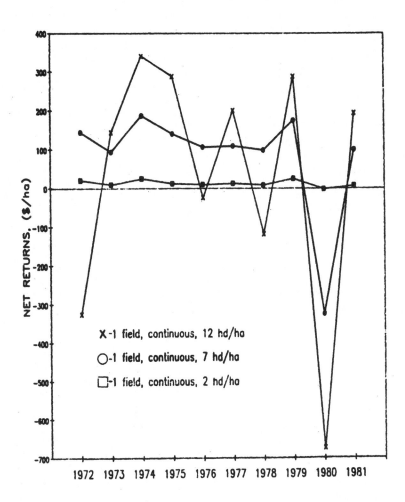

Figure 2. Net returns for three selected grazing
management strategies, Booneville, Ark.

1979), higher stocking rates result in higher net returns whereas in "bad" years (1980) the converse is true. In "intermediate" years (1976, 1978) different rankings surface.

The information in Figure 2 is important in that alternative decision makers will rank these three strategies differently depending on their preference for, or aversion to, risk. Given that a decision maker's attitude toward risk is known, a necessary piece of information in the choice of the preferred strategy is an assessment of the unknown probabilities and distribution characteristics underlying each strategy. An estimate of these distribution characteristics is obtained from the simulated model output by developing an empirical distribution function (Hogg and Tanis, 1977) for each strategy. Table 13 converts the simulated output of Figure 2 into empirical cumulative frequency distributions. As final output from the simulation model, these distributions can serve as the basis for choosing the preferred set of strategies under alternative risk-efficiency criteria.[4]

Table 13. Sample cumulative frequency distributions for 3 simulated strategies, $/Ha.

(1 Field, Continuous Grazing, 3 Stocking Rates)		
Strategy: (1, C, 2)	(1, C, 7)	(1, C, 12)
Cumulative Frequency[a]		
.10 −5.07	−333.83	−679.22
.20 15.55	85.12	−323.83
.30 15.76	90.19	−113.25
.40 15.89	94.99	−25.55
.50 18.48	109.68	143.73
.60 24.17	114.56	185.97
.70 29.24	139.19	206.00
.80 31.78	143.54	278.20
.90 37.47	163.74	292.24
1.00 40.73	177.05	333.02

[a]Probability of net returns being less than or equal to a given dollar value.

---

[4]Concepts of producer risk attitudes and alternative risk-efficiency models are described in Barry (1984).

SUMMARY

This chapter has provided an example of the role that the biological-phenological Kentucky Beef Forage Model GRAZE plays in economic analysis of grazing management systems. A computer experiment was designed which would enable assessment of

(a) grazing system performance under a variety of alternative manager-controlled grazing strategies; and,

(b) sensitivity of grazing system performance to weather variability for each of the strategies analyzed.

Economic performance of the GRAZE model output demonstrated that both high stocking rate (land intensive) and low stocking rate (animal intensive) grazing strategies result in lower expected net returns per hectare then did an intermediate stocking rate. The greatest mean net returns per hectare was achieved with a one-field, continuous grazing system stocked with 7 head per hectare.

Grazing management strategy performance was shown to be differentially sensitive to alternative "states of nature" represented by 10 years of historical weather data. Selected model output on economic performance was converted into empirical distribution functions in order to provide an assessment of the probability characteristics of the alternative management strategies. This latter form of model output is appropriate as a basis for use in determining preferred strategies under risk.

156

REFERENCES

Anderson, Jack R., John L. Dillon, and Brian Hardaker.
1977. Agricultural Decision Analysis. Iowa State
University Press, Ames, IA.

Barry, Peter J. (Ed.). 1984. Risk Management in Agriculture. Iowa State University Press, Ames, IA.

Beattie, Bruce R., and C. Robert Taylor. 1985. The
Economics of Production. John Wiley and Sons,
New York, NY.

Brown, Connell J. 1984. Personal communication.
Department of Animal Science, University of Arkansas.

Hogg, Robert V., and Elliot A Tanis. 1977. Probability and Statistical Inference, (Chapter 3). Macmillan Publishing Co., New York, NY.

Loewer, O.J., W. Butts, S.W. Coleman, L.L. Erlinger,
H.W. Essig, J.P. Fontenot, N. Gay, A.C. Linnerud,
C. Long, R. Muntifering, J.W. Oltjen, D.G. St.
Louis, J.A. Stuedemann, K. Taul, and L. Turner.
1985a. "The Animal Component." In Simulation of
Forage and Beef Production in the Southern Region,
Vance H. Watson and Chester M. Wells Jr., (Eds.).
Southern Cooperative Series, Bulletin 308.

Loewer, O.J., K.L. Taul, W. Turner, N. Gay, and
R. Muntifering. 1985b. "Modeling of Selective
Grazing of Beef Animals as Influenced by Environment." ASAE Paper No. 85-4008, St. Joseph, MI
49085.

Schlaifer, Robert. 1959. Probability and Statistics
for Business Decisions. McGraw-Hill Book Company,
New York, NY.

Smith, E.M., L.M. Tharel, M.A. Brown, G.W. Burton, C.T.
Dougherty, S.L. Fales, V.H. Watson, and G.A. Peder-
son. 1985. "The Plant Growth Component." In
Simulation of Forage and Beef Production in the
Southern Region. Vance H. Watson and Chester M.
Wells Jr. (Eds.). Southern Cooperative Series,
Bulletin 308.

Spaudling, B.W., W.A. Halbrook, C.J. Brown, C.R.
    Garner, D.C. Steiger, and J.A. Clower. 1983.
    "Feeder Cattle Production Budgets, 1982-1983."
    Special Report 109, AES-CES, University of Arkan-
    sas, Fayetteville, Arkansas.

USDA-AMS Agricultural Marketing Service. (Livestock,
    Poultry, Grain and Seed Division.) Livestock
    detailed annual quotations, Little Rock, Arkansas.

VII

The Texas A&M Beef Cattle Simulation Model

T.C. Cartwright and P.E. Doren

INTRODUCTION

Traditionally, application of animal science to the livestock production process has been based on three general assumptions. One is that increased productivity in any single component of the production process will result in greater net offtake of a production system. Another is that production units tend to respond similarly to inputs across geographical areas and types of production systems. The third is that biological efficiency of production is closely correlated with profitability.

The effect of accepting these assumptions as being essentially correct, if not perfect, is that our recommendations have tended to be simplistic and general, ignoring the network of interactions existing among genetic, environment and economic inputs and outputs.

Perhaps this approach is the reason that beef cattle production, evaluated with either biological or economic criteria, has not been enhanced to the extent that separate research achievements in animal science would lead us to expect. We now have the technology to identify and quantify these interactions and, to the extent that these interactions importantly affect offtake, to develop recommendations for each ecological area and type of production system with adjustments being continuously made on a current basis. That is to say, we have the hardware, but the software is just beginning to emerge.

Animal Science Department, Texas A&M University. Parts of this manuscript including the tables and figures are taken from Cartwright (1982), Kothmann and Smith (1983), and Doren, et al. (1985).

Our original concerns in systems analysis were oriented, as animal geneticists, around developing evaluation criteria (for either individuals or breeds) that would lead to selecting more productive cattle for the beef cattle production unit. This concern led us into applying systems analysis techniques and, therefore, into examining genotypic and environmental effects in concert rather than in isolation.

Nevertheless, the genotype of each individual beef animal mediates the expression of its performance and is therefore the basis upon which we founded the structure of our beef cattle model. Consideration of basic interactions among genetic potential and environment is necessary to understand the Texas A&M Beef Cattle Simulation Model.

BEEF CATTLE PRODUCTION SYSTEMS

A beef cattle production system may be divided into two phases: I. calf production (cow-calf), and II. weight increase (steers mostly); Phase I relates mostly to increasing numbers and Phase II relates to increasing weight and finish. Breeding and selection have been directed mostly toward increasing the output of one or both of these phases. Only minor attention has been given to reducing input in relation to output. Cows and heifers consume most of the resources related to Phase I and steers (or bulls) and excess heifers consume most of the resources related to Phase II but both overlap. Traits that have a major impact on Phase I are principally cow or maternal traits include low maintenance requirements; early puberty, high fertility, calving ease; moderate milk production; and productive longevity, adapted, hardy. Traits that have a major impact on Phase II include principally slaughter animal or paternal traits such as fast, efficient growth (low feed/gain) and high cut-out, meatiness, desirable carcass.

The traits that are important in Phase I are not all compatible with those traits that are important in Phase II. Obviously, for beef cattle, selection of a breed based solely on either the desired cow herd traits or desired meat animal traits is not logical since each includes only components of the total, complex production system. Net herd production (i.e. herd production after inputs are considered) tends to be enhanced by a balance among these traits. This antagonism which is complex and varies in degree from one production system to another may be a major deterrent to selecting cattle with the genetic potential for improving net herd production.

Nevertheless, selection criteria must be based on characters and practiced by keeping or rejecting each individual animal. Therefore we usually conceive of characters in terms of individuals, but when individuals are aggregated into production units, or herds, the plusses and minuses do not simply sum but rather interact in a complex manner. Objective, metric characters usually considered in beef cattle selection are designated for the convenience of man and are often biologically interrelated. These characters may be divided into two general types: primary and ancillary. We classify three characters as primary: size, maturing rate, and milk production. These primary characters have pervasive correlated effects; i.e., they affect or are affected by many other characters. Ancillary characters include reproductive soundness, calving ability, bone/muscle/fat ratio, structural soundness, adaptability, disease and parasite resistance, temperament, color, and horned condition.

The TAMU model is structured around the primary characters. Size is a composite character that is intended here to relate to structural size: it is conveniently characterized by body weight at maturity, at a given body composition. Generally, as genetic size potential increases, rate of gain potential increases and degree of maturity, including degree of finish at any age, decreases. Cow size is important because of effects on growth rate, maturing rate and weight, and therefore on feed requirements (or stocking rates) for maintenance, growth, and age at first calf. The nutrients consumed by cows are the major expense related to beef production. Cow size is also important because of the genetic potential for growth and maturing rate transmitted to her progeny. Level of milk production affects nutrient requirements, degree of fatness, and rebreeding of the cow as well as weaning weight and condition of her calf. Maturing rate, independent of size, affects age at puberty and degree of condition at any age and therefore age at first calf and rebreeding.

Size and milk production potentials can be changed relatively easily by selection (i.e., heritability is in the medium to high range). Maturing rate, independent of size (relative maturing rate in the animal science literature but more commonly, an absolute rate), is much more difficult to measure and to change by selection. Breeds exist that combine various size and milk production potentials and, to some degree, differences in relative maturing rate. Examples of contrasts are Charolais vs. Angus for size (currently many Angus overlap Charolais), Simmental vs. Hereford for milk, and Zebu vs. European breeds for maturing rate.

Selection of livestock may be considered as genetically changing an animal population so that it is better adapted to its environment. Enviroment (E) as used here is broadly defined to include all nongenetic effects both physical and economic, e.g., market prices and input costs are part of E effects. The production and economic environment is not static as sometimes assumed but is dynamic and fluctuates yearly. Therefore, breeding beef cattle, so that they are best adapted to their environment (i.e., so that net herd production is maximal), is a dynamic process where breeding goals are moving targets. At the same time, the environment is usually manipulated to various degrees to better fit the cattle. These changes tend to make breeding, in effect, an iterative process. Systems analysis is a useful approach for examining how well a breed, or some smaller breeding unit, is synchronized with its environment, and how it may be genetically changed to more closely approach continuous synchronization over the good years and poor years and in times of relatively high prices and low prices.

MODEL DESCRIPTION

The Texas A&M Cattle Production Systems Model (Sanders and Cartwright, 1979a,b) accounts for animals on the basis of classes which are determined by sex and age, and in the case of breeding age females, on the basis of month of lactation or gestation. The original model was modified and restructured to account for individual animals (D.W. Mayfield, unpublished manuscript, 1979; Baker, 1982).

The newer model, the Texas A&M Beef Cattle Simulation Model (Doren, et al., 1985), is a computer model programmed in FORTRAN IV. It has a top-down hierarchial organization of 50 subroutines. Animal records are handled in bidirectional lists with pointers set to designate herds or class/mating groups. Stochastic elements are associated with birth, death, estrus, conception, and removal.

Forrester (1968) outlines methods for defining systems through the identification of feedback loops; he stated:

"In concept a feedback system is a closed system. Its dynamic behavior arises within its internal structure. Any interaction which is essential to the behavior mode being investigated must be included inside the system boundary."

He reemphasized the importance of this component by stating that "The feedback loop is the basic structural element in systems. Dynamic behavior is generated by feedback." Figure 1 represents the feedback loops which are found in the model.

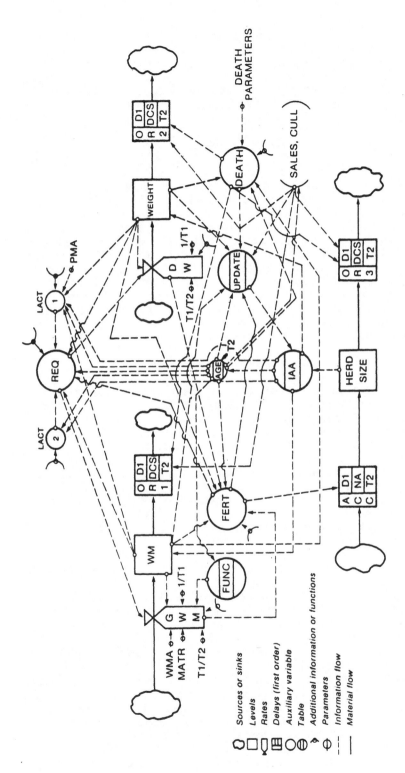

Figure 1. Major feedbacks in the Texas A&M Beef Cattle Simulation Model.

The levels, or integrations, in the model are weight, structural size (WM) and herd size. An example of a feedback loop is illustrated by the information flow from WM to the daily gain in WM (GWM), the rate that determines the amount of increase (of material) in WM for the month. This loop indicates that GWM is dependent upon WM(t) where:

$$GWM = f(WM(t))$$

and t is time in months.

However, GWM is also dependent upon the genetic potentials for mature size (WMA) and maturing rate (MATR), time constants (T1 = 1 d, T2 = 1 mo (30 d)), a tabular function (FUNC), and other adjustments. WM is represented by one of two functions used over distinct intervals which are dependent upon age (Sanders and Cartwright, 1979b). Therefore, FUNC is used to determine whether GWM is designated as a constant, implying a linear increase in WM and no direct dependence upon WM (i.e. the previously mentioned information flow from WM to GWM is 0), or if is a self-limiting rate which is proportional to the difference:

$$WMA - WM(t).$$

Even if GWM is a constant, a feedback exists from WM through nutritional requirements (REQ) to GWM. If the materials (TDN) required for GWM in an individual of size WM(t) are severely limited, then GWM may be reduced (representing stunted growth).

The time constants, T1 and T2, are necessary to maintain consistency of units in the equations and to determine the timing characteristics of the simulation. T1 is used as a constant to determine GWM and the daily gain in weight (DW), but its more subtle utility lies in designating the interval over which a given amount of TDN is converted to live weight gain. However, the timing of events in the model occur at 1 mo intervals, thus T2 is required to set the rates and delays on that time scale and to increment age (AGE).

Two auxiliary variables are used to determine the effects of lactation in the model; LACT1 representing the lactation of an individual and LACT2 representing the lactation of the dam of an individual. LACT1 is dependent upon a parameter (PMA) used as an index of the genetic potential for milk production, among other factors. PMA is defined as peak day lactation at maturity with no nutritional or other stress.

Sales and cull functions, as well as the auxiliary variable DEATH, determine removals while FERT determines additions to the herd. The sales and cull functions are the sources of much of the influence which external management practices (e.g., stochastic elements associated with age at culling, number of heifers retained, season of weaning, etc.) have in the model. FERT is also directly influenced by management practices (minimum age at exposure to breeding, months of breeding season) and not only increments the accumulation delay (ACC) but also represents the reproductive status of the individual. The value of DEATH is determined by stochastic elements which are dependent upon the month of the year, AGE and condition (WEIGHT/WM) of the animal, among other factors.

DEATH, FERT, and the sales and cull functions are unique in that their roles transcend levels of the hierarchy in the model. The output from these variables and functions are interpreted in both the individual and herd level dynamics. This interpretation is made possible through the use of tabled values of individual animal attributes that are contained in IAA. The contents of IAA are altered at the end of each month by the table UPDATE. The information flow from HERD SIZE to IAA does not constitute a pathway for a feedback through any of the components on the individual level of the hierarchy. The mechanism that limits HERD SIZE is the array size set for individual records in the computer program.

The delays in Figure 1, over a given time interval, T2, produce an output signal with an instantaneous rise and decay. ACC will accumulate the number of births (or incoming animals, NA) that occurred in a given month but will not "discharge" into HERD SIZE until the end of the current month of simulation. The same type of behavior is observed for the output rates (OR1, OR2 and OR3) but the influencing factors are deaths, culls and sales (DCS).

The cloud-shaped structures in Figure 1 represent either sources or sinks. A material which flows from a source to WM through GWM could be forage. The production of the forage or how it is made available to an individual animal is not a part of the system which this model simulates, therefore any forage input, or any other input to the system, is represented as arising from a source. By the same token, outputs from the system are placed in sinks.

The computer model requires input values that designate forage quality (digestibility and crude protein content) and availability on a monthly basis. These values do not change in response to any behavior of the system but influence behavior within the system. Management practices can be considered disturbances, since they are represented as parameters which are determined outside the feedback structure of the system.

The purpose of this model is to describe a system involving the growth, reproduction, and lactation of beef cattle. The information which is obtained from it

relates to the response behavior of the system to specified inputs or disturbances (genotype, management, and
feed resource).

REPORT SUMMARIES

The Texas A&M Beef Cattle Simulation Model, as it
is now constructed, makes a considerable amount of the
simulated data available to the user. These reports
vary in their degree of resolution from the amount
eaten by and physiological status of a particular animal to the total amount of forage consumed by a herd
over a twenty-year period. This report format allows
verification of the response behaviors to be obtained
from the simulations or the examinations of "bottom
line" information; whichever happens to be pertinent to
the user or situation.
Examples of some of the generated reports are
contained in Figures 2 through 14. All of these
reports are summaries; however, information on individuals may be obtained any time during execution by using
the "debugging" statements that are present within the
program. Although output is voluminous when the
"debugging" statements are invoked, this capability is
critical to validation and verification.
The report in Figure 2 represents what is referred
to as the monthly summary. At the beginning of each
simulated month the upper portion of the report is
generated following passes made through the "biological" subroutines. The statistics are collected on the
basis of herd, pasture, class and age groups. The
management report indicates transfers made from classes
or pastures, births and removals due to deaths, sales,
or culls.
Some examples of certain types of animal performance information that can be obtained are presented in
Figures 3 and 4. Both reports reveal measures of calf
performance relative to the month of birth, sex of the
calf and age of the dam.
The yearly summary (Figure 5) involves only a
small amount of data when compared to the monthly
summary because statistics are printed by herd, pasture
and class groups for each month. However, the ultimate
in data reduction is the run summary (Figure 6) which
tallies overall measures for specified years.
Figure 7 through 12 present reports generated from
a set of SAS routines that are placed at the rear of
the FORTRAN program. Since we have begun to
concentrate our programming efforts on our own minicomputer we have lost the ability to use SAS; therefore
routines have been written to replace the reports that
were generated by that package.

39 1 APR BULL CAL 261   0   1   0   0   0

85 BIRTHS   5 DEATHS   0 SALES IN MONTH 4

STARTING YEAR 39 MONTH 5

--- AFTER 179 MONTHS ---

YR	HD	PA	MON	CLASS	AG GR	NUM	DTH	STL BIR	BIR	AGE	AVE	SD	AVE	SD	TN	TON	ADG	CYC PREG X	NO.	PREG X	NO.	MILK PROD AVE	SD	% PREG
39	1	1	MAY	SALE HRF	5	3	0	0	0	4.0	89	7.4	89	3.6	1.8	1.0	0.348	0	0	0	74	0.0 0.0		0.0
39	1		MAY	SALE HRF		3	0	0	0	4.0	89	7.	89	4.	1.8	1.0	0.348	0	0	0	83	0.0 0.0		0.0
39	1	1	MAY	BR COWS	8	92	0	0	0	27.3	382	17.4	403	5.4	8.4	5.2	0.166	16	33	74	6.0 0.69		35.87
39	1	1	MAY	BR COWS	10	92	0	0	0	39.3	399	26.8	450	8.6	10.8	6.1	0.172	14	24	63	5.9 0.87		26.09
39	1	1	MAY	BR COWS	12	88	0	0	0	51.2	416	24.5	471	6.6	11.3	6.5	0.225	13	30	53	6.1 0.86		34.09
39	1	1	MAY	BR COWS	13	78	0	0	0	63.2	432	27.2	485	4.5	11.6	6.6	0.245	15	22	56	6.8 1.16		28.21
39	1	1	MAY	BR COWS	14	76	0	0	0	75.1	445	23.1	492	1.7	11.9	6.6	0.187	15	22	59	7.4 1.08		28.95
39	1	1	MAY	BR COWS	15	80	0	0	0	87.3	440	27.0	496	1.5	11.8	6.7	0.231	26	20	61	6.9 1.23		25.00
39	1	1	MAY	BR COWS	16	71	0	0	0	99.0	451	25.1	498	1.2	11.5	6.6	0.256	22	28	47	7.3 1.04		40.85
39	1	1	MAY	BR COWS	17	74	0	0	0	111.0	446	26.3	499	0.8	11.5	6.4	0.239	24	20	49	6.8 1.26		27.03
39	1	1	MAY	BR COWS	18	74	0	0	0	123.3	441	23.8	499	0.6	11.2	6.3	0.181	17	22	55	6.5 1.10		29.73
39	1		MAY	BR COWS		725	0	0	0	72.6	426	34.	475	32.	11.2	6.3	0.222	16	222	537	6.6 1.15		30.62
39	1	2	MAY	RPL HFRS	14	96	0	0	0	15.2	296	14.4	286	14.5	6.3	3.5	0.558	5	12	0	0.0 0.0		12.50
39	1		MAY	RPL HFRS		96	0	0	0	15.2	296	14.	286	14.	6.3	3.5	0.558	5	12	0	0.0 0.0		12.50
39	1	1	MAY	HFR CALF	2	48	0	0	0	1.0	54	1.7	46	0.1	0.3	0.2	0.862	0	0	0	0.0 0.0		0.0
39	1	1	MAY	HFR CALF	3	54	0	0	0	2.0	77	3.0	64	0.1	0.5	0.4	0.732	0	0	0	0.0 0.0		0.0
39	1	1	MAY	HFR CALF	4	71	0	0	0	3.0	87	4.7	82	0.4	0.6	0.4	0.737	0	0	0	0.0 0.0		0.0
39	1	1	MAY	HFR CALF	5	102	0	0	0	4.0	115	8.6	99	3.8	0.6	0.5	0.695	0	0	0	0.0 0.0		0.0
39	1		MAY	HFR CALF		276	0	0	0	2.8	92	23.	78	20.	0.6	0.3	0.755	0	0	0	0.0 0.0		0.0
39	1	1	MAY	FATN BUL	5	2	0	0	0	4.0	99	26.3	96	21.1	2.1	1.2	0.228	0	0	0	0.0 0.0		0.0
39	1		MAY	FATN BUL		2	0	0	0	4.0	99	26.	96	21.	2.1	1.2	0.228	0	0	0	0.0 0.0		0.0
39	1	1	MAY	BULL CAL	2	35	1	0	0	1.0	90	1.9	52	0.1	0.3	0.2	0.980	0	0	0	0.0 0.0		0.0
39	1	1	MAY	BULL CAL	3	54	0	0	0	2.0	84	3.1	73	0.1	0.5	0.3	0.886	0	0	0	0.0 0.0		0.0
39	1	1	MAY	BULL CAL	4	60	1	0	0	3.0	107	4.5	94	0.5	0.7	0.4	0.831	0	0	0	0.0 0.0		0.0
39	1	1	MAY	BULL CAL	5	112	0	0	1	4.0	126	9.8	114	5.8	0.9	0.5	0.776	0	0	0	0.0 0.0		0.0
39	1		MAY	BULL CAL		261	3	0	0	3.0	104	25.	93	23.	0.7	0.4	0.840	0	0	0	0.0 0.0		0.0

*** MANAGEMENT REPORT FOR MAY YEAR 39 ***

YR	HD	MON	CLASS	NUM	TRANSF IN	OUT	SALES NUM	WEIGHT	WM
39	1	MAY	SALE HRF	3	0	0	0	0	0
39	1	MAY	SALE COW	3	0	0	0	0	0
39	1	MAY	BR COWS	725	0	0	0	0	0
39	1	MAY	RPL HFRS	96	0	0	0	0	0
39	1	MAY	HFR CALF	276	0	0	0	0	0
39	1	MAY	OXEN	0	0	0	0	0	0
39	1	MAY	STEERS	0	0	0	0	0	0
39	1	MAY	STR CALF	0	0	0	0	0	0
39	1	MAY	FATN BUL	2	0	0	0	0	0
39	1	MAY	BR BULLS	0	0	0	0	0	0
39	1	MAY	REPL BUL	0	0	0	0	0	0
39	1	MAY	BULL CAL	258	0	0	0	0	0

0 BIRTHS   3 DEATHS   0 SALES IN MONTH 5

STARTING YEAR 39 MONTH 6

--- AFTER 180 MONTHS ---

Figure 2. Monthly Summary generated by the Texas A&M Beef Cattle
Simulation Model

168

Figure 3. Summary of calf weights by month, month of birth, sex and age of dam

*** CALF SURVIVAL BY COW AGE, SEX AND BIRTH MONTH ***

BIR MON	MON	S	X	2.0	2.5	3.0	3.5	4.0	5.0	6.0	7.0	8.0	9.0	10.0	11.0	12.0	13.0	14.0	15.0	AVE
FEB	FEB	M	NO	9	0	17	0	14	14	12	11	11	13	12	0	0	0	0	0	112
			%	1.00	0.0	1.00	0.0	1.00	1.00	1.00	1.00	1.00	1.00	1.00	0.0	0.0	0.0	0.0	0.0	1.00
		F	NO	7	0	17	0	12	13	12	14	12	11	10	0	0	0	0	0	107
			%	1.00	0.0	1.00	0.0	1.00	1.00	1.00	1.00	1.00	1.00	1.00	0.0	0.0	0.0	0.0	0.0	1.00
			NO	16	0	33	0	26	26	24	25	23	23	22	0	0	0	0	0	219
			%	1.00	0.0	1.00	0.0	1.00	1.00	1.00	1.00	1.00	1.00	1.00	0.0	0.0	0.0	0.0	0.0	1.00
FEB	MAR	M	NO	9	0	16	0	13	13	12	11	11	12	11	0	0	0	0	0	107
			%	0.95	0.0	0.97	0.0	0.95	0.94	0.96	0.94	0.94	0.95	0.93	0.0	0.0	0.0	0.0	0.0	0.95
		F	NO	6	0	16	0	12	12	12	13	12	10	10	0	0	0	0	0	103
			%	0.93	0.0	0.96	0.0	0.94	0.96	0.96	0.96	0.95	0.98	0.96	0.0	0.0	0.0	0.0	0.0	0.96
			NO	15	0	32	0	24	25	23	24	22	23	21	0	0	0	0	0	209
			%	0.94	0.0	0.97	0.0	0.94	0.95	0.96	0.95	0.95	0.96	0.95	0.0	0.0	0.0	0.0	0.0	0.95
FEB	APR	M	NO	8	0	16	0	12	13	12	11	11	12	11	0	0	0	0	0	106
			%	0.94	0.0	0.96	0.0	0.92	0.93	0.96	0.94	0.94	0.95	0.93	0.0	0.0	0.0	0.0	0.0	0.94
		F	NO	6	0	16	0	11	12	12	13	11	10	10	0	0	0	0	0	102
			%	0.93	0.0	0.96	0.0	0.93	0.96	0.96	0.96	0.93	0.98	0.96	0.0	0.0	0.0	0.0	0.0	0.95
			NO	15	0	32	0	24	25	23	24	22	23	21	0	0	0	0	0	208
			%	0.94	0.0	0.96	0.0	0.92	0.95	0.96	0.95	0.93	0.96	0.95	0.0	0.0	0.0	0.0	0.0	0.95
FEB	MAY	M	NO	8	0	16	0	12	13	11	11	10	12	11	0	0	0	0	0	105
			%	0.94	0.0	0.96	0.0	0.91	0.93	0.95	0.94	0.93	0.94	0.93	0.0	0.0	0.0	0.0	0.0	0.94
		F	NO	6	0	16	0	11	12	12	13	11	10	10	0	0	0	0	0	102
			%	0.93	0.0	0.96	0.0	0.92	0.96	0.96	0.96	0.92	0.98	0.96	0.0	0.0	0.0	0.0	0.0	0.95
			NO	15	0	32	0	24	25	23	24	22	22	21	0	0	0	0	0	207
			%	0.94	0.0	0.96	0.0	0.92	0.94	0.95	0.95	0.93	0.96	0.94	0.0	0.0	0.0	0.0	0.0	0.94
FEB	JUN	M	NO	8	0	16	0	12	13	11	11	10	12	11	0	0	0	0	0	105
			%	0.94	0.0	0.96	0.0	0.91	0.93	0.94	0.94	0.93	0.94	0.93	0.0	0.0	0.0	0.0	0.0	0.94
		F	NO	6	0	16	0	11	12	12	13	11	10	10	0	0	0	0	0	102
			%	0.93	0.0	0.96	0.0	0.92	0.96	0.96	0.96	0.92	0.98	0.96	0.0	0.0	0.0	0.0	0.0	0.95
			NO	15	0	32	0	24	25	23	24	22	22	21	0	0	0	0	0	207
			%	0.94	0.0	0.96	0.0	0.92	0.94	0.95	0.95	0.93	0.96	0.94	0.0	0.0	0.0	0.0	0.0	0.94
FEB	JUL	M	NO	8	0	16	0	12	13	11	11	10	12	11	0	0	0	0	0	105
			%	0.94	0.0	0.96	0.0	0.91	0.93	0.94	0.93	0.93	0.93	0.93	0.0	0.0	0.0	0.0	0.0	0.93
		F	NO	6	0	16	0	11	12	12	13	11	10	10	0	0	0	0	0	102
			%	0.93	0.0	0.96	0.0	0.92	0.96	0.96	0.96	0.92	0.98	0.96	0.0	0.0	0.0	0.0	0.0	0.95
			NO	15	0	32	0	24	25	23	24	22	22	21	0	0	0	0	0	206
			%	0.94	0.0	0.96	0.0	0.91	0.94	0.95	0.95	0.93	0.95	0.94	0.0	0.0	0.0	0.0	0.0	0.94
FEB	AUG	M	NO	8	0	14	0	10	11	10	9	10	10	11	0	0	0	0	0	94
			%	0.87	0.0	0.86	0.0	0.77	0.83	0.87	0.82	0.86	0.82	0.88	0.0	0.0	0.0	0.0	0.0	0.84
		F	NO	6	0	15	0	11	11	10	12	10	10	9	0	0	0	0	0	93
			%	0.85	0.0	0.88	0.0	0.86	0.88	0.85	0.88	0.81	0.90	0.89	0.0	0.0	0.0	0.0	0.0	0.87
			NO	13	0	29	0	21	22	21	22	20	20	20	0	0	0	0	0	187
			%	0.86	0.0	0.87	0.0	0.81	0.86	0.86	0.86	0.84	0.86	0.88	0.0	0.0	0.0	0.0	0.0	0.85
FEB	SEP	M	NO	8	0	14	0	10	11	10	9	10	10	11	0	0	0	0	0	94
			%	0.86	0.0	0.86	0.0	0.77	0.83	0.87	0.82	0.86	0.82	0.88	0.0	0.0	0.0	0.0	0.0	0.84
		F	NO	6	0	15	0	11	11	10	12	10	10	9	0	0	0	0	0	93
			%	0.85	0.0	0.88	0.0	0.86	0.88	0.85	0.88	0.81	0.90	0.89	0.0	0.0	0.0	0.0	0.0	0.87
			NO	13	0	29	0	21	22	21	22	20	20	20	0	0	0	0	0	187
			%	0.85	0.0	0.87	0.0	0.81	0.85	0.86	0.86	0.84	0.86	0.88	0.0	0.0	0.0	0.0	0.0	0.85
FEB	OCT	M	NO	8	0	14	0	10	11	10	9	10	10	11	0	0	0	0	0	94

Figure 4. Report summary of calf survival by month, month of birth, sex and age of dam

*** YEARLY SUMMARY ***

YR	HD	PA	MON	CLASS	AG GR	NUM	DTH	STL BIR	BIR	AGE	---W--- AVE	SD	---WN--- AVE	SD	TN	TDN	ADG	CYC PREG %	NO.	--MILK PROD-- NO.	AVE	SD	% PREG
24	1		JUL	BR COWS		670	0	0	0	69.4	433	37	476	28	10.2	5.3	0.094	6	464	507	5.5	0.84	69.25
24	1		JUL	RPL HFRS		98	0	0	0	17.1	321	16	312	14	6.6	3.4	0.469	5	59	0	0.0	0.0	60.20
24	1		JUL	HFR CALF		246	0	0	0	4.9	134	21	115	19	1.1	0.6	0.658	0	0	0	0.0	0.0	0.0
24	1		JUL	FATN BUL		1	0	0	0	79	80	0.	0.	1.8	0.6	0.0	0	0	0	0.0	0.0	
24	1		JUL	BULL CAL		261	1	0	0	4.9	150	23	134	22	1.5	0.8	0.764	0	0	0	0.0	0.0	
24	1		AUG	BR COWS		670	0	0	0	70.1	477	39	477	21	9.8	4.3	0.052	4	514	507	4.9	0.73	76.72
24	1		AUG	RPL HFRS		98	0	0	0	18.1	334	16	325	13	6.1	3.3	0.423	4	77	0	0.0	0.0	78.57
24	1		AUG	HFR CALF		246	0	0	0	5.9	153	21	133	19	1.7	0.9	0.639	0	0	0	0.0	0.0	0.0
24	1		AUG	FATN BUL		1	0	0	0	5.9	79	0.	80	0.	1.7	0.7	0.0	0	0	0	0.0	0.0	
24	1		AUG	BULL CAL		261	2	0	0	7.0	172	23	155	22	2.9	1.5	0.750	0	0	0	0.0	0.0	
24	1		SEP	SALE HRF		1	0	0	0	7.0	160	0.	150	0.	2.9	1.5	0.340	0	0	504	4.3	0.56	0.0
24	1		SEP	BR COWS		669	0	0	0	71.5	437	41	478	25	9.8	4.9	0.080	9	513	504	4.5	0.0	76.68
24	1		SEP	RPL HFRS		97	0	0	0	19.1	347	16	337	12	6.3	3.5	0.434	4	77	0	0.0	0.0	79.38
24	1		SEP	HFR CALF		245	0	1	0	6.9	172	21	151	19	2.3	1.1	0.634	0	0	0	0.0	0.0	0.0
24	1		SEP	FATN BUL		1	0	0	0	8.0	79	0.	80	0.	1.7	0.9	0.0	0	0	0	0.0	0.0	
24	1		SEP	BULL CAL		259	0	0	0	6.9	195	24	176	22	2.8	1.4	0.748	12	513	0	0.0	0.0	76.68
24	1		OCT	BR COWS		669	1	0	0	72.5	447	40	480	24	9.4	4.5	0.353	7	77	0	0.0	0.0	39.09
24	1		OCT	RPL HFRS		197	0	0	0	14.0	274	87	258	90	5.3	2.7	0.425	14	508	0	0.0	0.0	77.09
24	1		NOV	BR COWS		659	1	0	0	72.8	286	81	280	23	3.9	3.1	0.220	9	77	0	0.0	0.0	39.29
24	1		NOV	RPL HFRS		196	0	0	0	15.0	284	85	271	27	5.1	2.6	0.413	15	507	0	0.0	0.0	77.05
24	1		DEC	BR COWS		658	1	0	0	73.8	297	83	481	42	4.9	2.8	0.138	9	77	0	0.0	0.0	39.29
25	1		DEC	RPL HFRS		196	5	4	210	16.0	293	51	283	61	3.5	2.4	0.359	17	583	0	0.0	0.0	77.32
25	1		JAN	BR COWS		754	8	0	0	68.1	459	18	468	41	3.5	3.3	0.392	0	0	0	0.0	0.0	0.0
25	1		JAN	RPL HFRS		98	1	0	0	11.1	229	0.	218	17	0.0	2.0	0.379	0	0	0	0.0	0.0	0.0
25	1		JAN	HFR CALF		97	0	0	0	0.0	29	0.	29	0.	0.0	0.0	0.0	0	0	0	0.0	0.0	
25	1		JAN	BULL CAL		109	0	0	0	0.0	30	0.	30	0.	0.0	0.0	0.0	0	0	0	0.0	0.0	
26	1		JAN	BR COWS		761	8	2	223	68.3	459	52	468	41	3.3	3.0	0.398	15	596	0	0.0	0.0	78.32
26	1		JAN	RPL HFRS		100	0	0	0	11.0	227	16	216	17	3.5	2.3	0.382	0	0	0	0.0	0.0	0.0
26	1		JAN	HFR CALF		116	0	0	0	0.0	29	0.	29	0.	0.0	0.0	0.0	0	0	0	0.0	0.0	
27	1		JAN	BR COWS		764	8	1	210	68.4	457	53	468	41	5.9	3.5	0.385	16	585	0	0.0	0.0	76.57
27	1		JAN	RPL HFRS		100	3	0	0	11.2	230	15	220	15	3.5	2.3	0.375	0	0	0	0.0	0.0	0.0
27	1		JAN	HFR CALF		101	0	0	0	0.0	29	0.	29	0.	0.0	0.0	0.0	0	0	0	0.0	0.0	
28	1		JAN	BR COWS		754	9	3	193	68.2	458	52	468	41	5.9	3.5	0.261	17	587	0	0.0	0.0	77.85
28	1		JAN	RPL HFRS		99	0	0	0	11.1	227	14	218	14	3.5	2.3	0.379	0	0	0	0.0	0.0	0.0
28	1		JAN	HFR CALF		109	0	0	0	0.0	29	0.	29	0.	0.0	0.0	0.0	0	0	0	0.0	0.0	
28	1		JAN	BULL CAL		81	0	0	0	0.0	30	0.	30	0.	0.0	0.0	0.0	0	0	0	0.0	0.0	
29	1		JAN	BR COWS		749	8	4	219	67.8	460	52	468	41	3.5	3.5	0.399	18	561	0	0.0	0.0	77.57
29	1		JAN	RPL HFRS		100	1	0	0	11.0	226	17	216	17	3.5	2.3	0.382	0	0	0	0.0	0.0	0.0
29	1		JAN	HFR CALF		116	0	0	0	0.0	29	0.	29	0.	0.0	0.0	0.0	0	0	0	0.0	0.0	
30	1		JAN	BULL CAL		99	0	3	222	67.9	459	52	468	42	5.9	3.5	0.395	19	572	0	0.0	0.0	76.68
30	1		JAN	BR COWS		102	0	0	0	11.2	229	15	220	15	3.5	2.3	0.375	0	0	0	0.0	0.0	0.0
31	1		JAN	BR COWS		743	7	2	204	68.3	459	52	468	41	5.9	3.5	0.384	18	571	0	0.0	0.0	76.85
31	1		JAN	RPL HFRS		100	2	0	0	11.0	226	17	216	17	3.5	2.3	0.382	0	0	0	0.0	0.0	0.0
31	1		JAN	HFR CALF		100	0	0	0	0.0	29	0.	29	0.	0.0	0.0	0.0	0	0	0	0.0	0.0	
32	1		JAN	BR COWS		741	6	2	222	68.2	461	52	468	41	5.9	3.5	0.402	16	584	0	0.0	0.0	78.81
32	1		JAN	RPL HFRS		99	1	0	0	11.0	226	17	216	17	3.5	2.3	0.381	0	0	0	0.0	0.0	0.0
32	1		JAN	HFR CALF		110	0	0	0	0.0	29	0.	29	0.	0.0	0.0	0.0	0	0	0	0.0	0.0	
33	1		JAN	BR COWS		747	5	4	213	68.2	459	54	468	41	5.9	3.5	0.388	18	576	0	0.0	0.0	77.11

Figure 5. Report summary of monthly totals collected in the yearly Summary

*** RUN SUMMARY ***

	STL			---CONSUMPTION---		--- MILK PRODUCTION ---	-------SALES-------		
YR	DTHS	BIR	BIR	TN	TDN	AVE DAILY	TOTAL W	WM	NO.
1	10	0	0	43939	21306	5.0	79318	72161	414
11	44	12	567	85551	45110	5.6	112061	107456	507
12	50	9	592	85859	45261	5.6	115643	110282	532
13	51	11	582	85690	45183	5.5	114572	110045	523
14	63	7	563	84770	44715	5.5	114156	109701	508
15	49	6	559	84346	44471	5.5	109232	104655	498
16	58	7	566	84540	44583	5.6	114580	110184	509
17	46	8	587	84417	44508	5.6	117678	112263	535
18	49	4	556	83773	44172	5.5	108000	103624	499
19	45	9	560	84142	44373	5.6	113093	108969	507
20	51	11	559	83712	44127	5.5	107473	102896	491

Figure 6. Report of annual totals collected in the Run Summary.

FEB MAY CALVING WITH WEANING AT END OF SEPT   1:33 WEDNESDAY, FEBRUARY 19, 1986

TABLE OF IP BY IAG

IP   IAG

FREQUENCY PERCENT ROW PCT COL PCT	7	8	10	12	13	14	15	16	17	18	TOTAL
0	55 0.37 1.66 63.22	384 2.57 11.61 20.79	219 1.47 6.62 11.77	494 3.31 14.93 27.94	418 2.80 12.64 24.62	392 2.63 11.85 23.99	334 2.24 10.10 21.13	342 2.29 10.34 22.24	347 2.33 10.49 23.51	323 2.17 9.76 22.62	3308 22.18
6	15 0.10 0.94 17.24	367 2.46 22.99 19.87	165 1.11 10.34 8.87	174 1.17 10.90 9.84	147 0.99 9.21 8.66	155 1.04 9.71 9.49	142 0.95 8.90 8.98	161 1.08 10.09 10.47	146 0.98 9.15 9.89	124 0.83 7.77 8.68	1596 10.70
7	17 0.11 0.70 19.54	403 2.70 16.52 21.82	375 2.51 15.38 20.16	242 1.62 9.92 13.69	271 1.82 11.11 15.96	224 1.50 9.18 13.71	242 1.62 9.92 15.31	228 1.53 9.35 14.82	208 1.39 8.53 14.09	229 1.54 9.39 16.04	2439 16.35
8	0 0.00 0.00 0.00	398 2.67 12.59 21.55	438 2.94 13.86 23.55	344 2.31 10.89 19.46	330 2.21 10.44 19.43	366 2.45 11.58 22.40	342 2.29 10.82 21.63	334 2.24 10.57 21.72	319 2.14 10.09 21.61	289 1.94 9.15 20.24	3160 21.18
9	0 0.00 0.00 0.00	295 1.98 6.68 15.97	663 4.44 15.02 35.65	514 3.45 11.64 29.07	532 3.57 12.05 31.33	497 3.33 11.26 30.42	521 3.49 11.80 32.95	473 3.17 10.72 30.75	456 3.06 10.33 30.89	463 3.10 10.49 32.42	4414 29.59
TOTAL	87 0.58	1847 12.38	1860 12.47	1768 11.85	1698 11.38	1634 10.95	1581 10.60	1538 10.31	1476 9.89	1428 9.57	14917 100.00

Figure 7. Table produced by SAS routine of month of gestation by age group observed in the month calving is initiated for the last ten years of simulation.

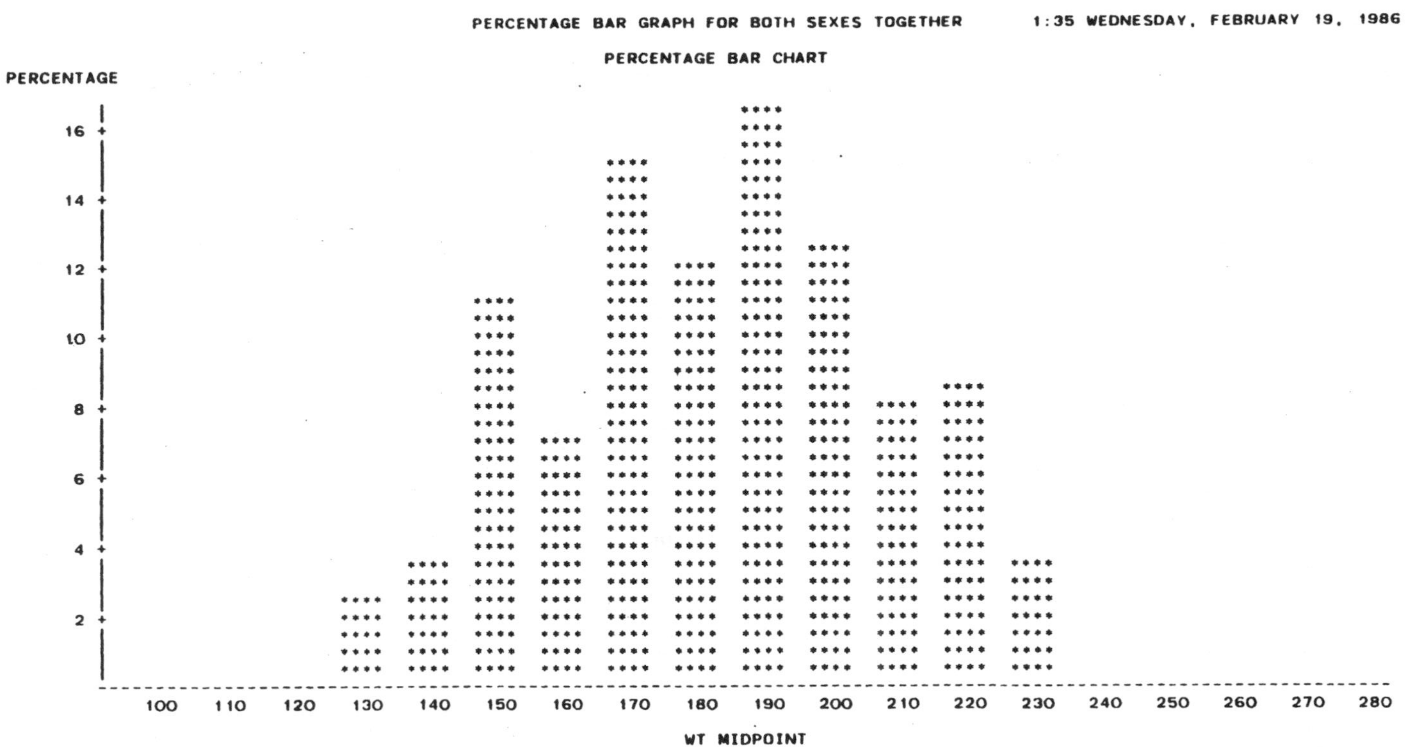

Figure 8. Distribution of weaning weights observed over the last ten years simulated. Plot generated by SAS routine.

174

SAS

TABLE OF CLASS BY MONTH

CLASS   MONTH

FREQUENCY PERCENT ROW PCT COL PCT	9	10	TOTAL
1	33 0.65 100.00 0.75	0 0.00 0.00 0.00	33 0.65
3	0 0.00 0.00 0.00	710 13.90 100.00 100.00	710 13.90
5	1629 31.88 100.00 37.03	0 0.00 0.00 0.00	1629 31.88
9	27 0.53 100.00 0.61	0 0.00 0.00 0.00	27 0.53
12	2710 53.04 100.00 61.60	0 0.00 0.00 0.00	2710 53.04
TOTAL	4399 86.10	710 13.90	5109 100.00

Figure 9. Table of number of individuals sold by class and month of sale for the last ten years simulated.

SAS   1:35 WEDNESDAY, FEBRUARY 19, 1986   3

VARIABLE	N	MEAN	STANDARD DEVIATION	MINIMUM VALUE	MAXIMUM VALUE	STD ERROR OF MEAN	SUM	VARIANCE	C.V.
--------- CLASS=1 ---------
WT	33	130.77636364	39.28759981	28.53000000	170.67000000	6.83909328	4315.6200000	1543.5154989	30.042
WM	33	127.53939394	35.16392869	28.53000000	165.01000000	6.12125428	4208.8000000	1236.5018809	27.571
--------- CLASS=3 ---------
WT	710	451.59978873	34.53500128	380.20000000	536.52000000	1.29607550	320635.85000	1192.6663135	7.647
WM	710	498.74477465	5.97064442	424.63000000	499.85000000	0.22407429	354108.79000	35.6485948	1.197
--------- CLASS=5 ---------
WT	1629	168.16205034	21.64331591	126.69000000	208.57000000	0.53624500	273935.98000	468.43312368	12.871
WM	1629	147.68597299	20.36930051	113.24000000	171.36000000	0.50467939	240580.45000	414.90840318	13.792
--------- CLASS=9 ---------
WT	27	125.44259259	57.79863011	30.29000000	216.70000000	11.12335155	3386.9500000	3340.6816430	46.076
WM	27	125.77592593	56.30525862	30.29000000	197.11000000	10.83595207	3395.9500000	3170.2821481	44.766
--------- CLASS=12 ---------
WT	2710	193.43964207	23.43752503	141.14000000	233.91000000	0.45022240	524221.43000	549.31757938	12.116
WM	2710	176.30714022	22.45015303	136.83000000	200.76000000	0.43125551	477792.35000	504.00937104	12.734

Figure 10. Statistics collected on the amounts of weight and WM sold during the last ten years of simulation.

FEB MAY CALVING WITH WEANING AT END OF SEPT   1:33 WEDNESDAY, FEBRUARY 19, 1986   4

TABLE 1 OF IAG BY LAC
CONTROLLING FOR PG=0

IAG   LAC

FREQUENCY PERCENT ROW PCT COL PCT	12	18	19	20	21	TOTAL
14	0 0.00 0.00 0.00	274 8.28 69.90 13.87	109 3.30 27.81 13.09	7 0.21 1.79 18.92	2 0.06 0.51 9.52	392 11.85
15	0 0.00 0.00 0.00	216 6.53 64.67 10.93	108 3.26 32.34 12.97	8 0.24 2.40 21.62	2 0.06 0.60 9.52	334 10.10
16	0 0.00 0.00 0.00	221 6.68 64.62 11.18	111 3.36 32.46 13.33	3 0.09 0.88 8.11	7 0.21 2.05 33.33	342 10.34
17	0 0.00 0.00 0.00	232 7.01 66.86 11.74	106 3.20 30.55 12.73	7 0.21 2.02 18.92	2 0.06 0.58 9.52	347 10.49
18	0 0.00 0.00 0.00	221 6.68 68.42 11.18	96 2.90 29.72 11.52	4 0.12 1.24 10.81	2 0.06 0.62 9.52	323 9.76
TOTAL	441 13.33	1976 59.73	833 25.18	37 1.12	21 0.63	3308 100.00

Figure 11. Table produced by SAS routine of age group by time since calving for nonpregnant cows and heifers one month prior to the initiation of calving in each of the last ten years simulated.

FEB MAY CALVING WITH WEANING AT END OF SEPT   1:33 WEDNESDAY, FEBRUARY 19, 1986   12

VARIABLE	N	MEAN	STANDARD DEVIATION	MINIMUM VALUE	MAXIMUM VALUE	STD ERROR OF MEAN	SUM	VARIANCE	C.V.
				IAG=14 MON=8					
IW	878	448.003	28.901	403.000	532.000	0.875	393347.000	835.292	6.451
IWM	878	493.334	1.306	487.000	496.000	0.044	433147.000	1.705	0.265
				IAG=14 MON=9					
IW	876	450.523	32.430	403.000	541.000	1.096	394658.000	1051.676	7.198
IWM	876	493.691	1.202	488.000	497.000	0.041	432473.000	1.446	0.244
				IAG=14 MON=10					
IW	876	461.264	29.695	414.000	550.000	1.003	404067.000	881.806	6.438
IWM	876	494.029	0.922	490.000	496.000	0.031	432769.000	0.851	0.187
				IAG=14 MON=11					
IW	872	467.611	31.480	419.000	559.000	1.066	407757.000	990.966	6.732
IWM	872	494.390	0.901	489.000	497.000	0.031	431108.000	0.812	0.182
				IAG=14 MON=12					
IW	869	471.603	34.422	420.000	568.000	1.168	409823.000	1184.896	7.299
IWM	869	494.632	0.742	491.000	497.000	0.025	429835.000	0.551	0.150
				IAG=15 MON=1					
IW	1581	483.908	40.414	415.000	567.000	1.016	765058.000	1633.275	8.352
IWM	1581	494.960	0.889	490.000	497.000	0.022	782531.000	0.789	0.180
				IAG=15 MON=2					
IW	861	465.215	33.373	422.000	572.000	1.137	400550.000	1113.787	7.174
IWM	861	495.210	0.792	491.000	498.000	0.027	426376.000	0.627	0.160
				IAG=15 MON=3					
IW	855	451.649	30.622	407.000	557.000	1.047	386160.000	937.736	6.780
IWM	855	495.414	0.665	492.000	498.000	0.023	423578.000	0.442	0.134
				IAG=15 MON=4					
IW	854	442.117	25.416	417.000	561.000	0.870	377568.000	645.951	5.749
IWM	854	495.570	0.553	494.000	497.000	0.019	423217.000	0.306	0.112
				IAG=15 MON=5					
IW	852	442.930	17.149	397.000	519.000	0.588	377376.000	294.082	3.872
IWM	852	495.857	0.588	493.000	498.000	0.020	422470.000	0.346	0.119
				IAG=15 MON=6					
IW	851	446.340	22.340	400.000	528.000	0.766	379835.000	499.067	5.005
IWM	851	496.071	0.665	493.000	498.000	0.023	422156.000	0.442	0.134

Figure 12.   Statistics collected on weight and WM by age group and month for last ten years simulated

```
FEED CONSUMPTION FOR HERD 1 AVERAGED ACROSS YEARS

YR HERD MON   CLASS  PAST   DRY MATTER   DIG. D.M.
----------------------------------------------------


   AV   1   JAN TOTAL       1        4364.        2619.
   AV   1   JAN TOTAL       2         348.         227.
   AV   1   JAN TOTAL      13        4712.        2845.




   AV   1   FEB BRD COWS    1        4325.        2595.
   AV   1   FEB BRD COWS   13        4325.        2595.

   AV   1   FEB REPL HFR    2         349.         227.
   AV   1   FEB REPL HFR   13         349.         227.

   AV   1   FEB HFR CALF    1          31.          19.
   AV   1   FEB HFR CALF   13          31.          19.




   AV   1   FEB BULL CLF    1          35.          21.
   AV   1   FEB BULL CLF   13          35.          21.

   AV   1   FEB TOTAL       1        4390.        2634.
   AV   1   FEB TOTAL       2         349.         227.
   AV   1   FEB TOTAL      13        4738.        2861.


   AV   1   MAR SALE HFR    1           1.           1.
   AV   1   MAR SALE HFR   13           1.           1.


   AV   1   MAR BRD COWS    1        4280.        2568.
   AV   1   MAR BRD COWS   13        4280.        2568.

   AV   1   MAR REPL HFR    2         507.         330.
   AV   1   MAR REPL HFR   13         507.         330.
```

Figure 13. Summary of feed consumption by month, herd, pasture
and class averaged over the last ten years simulated.

**** NUMBER OF CATTLE IN HERD 1 AVERAGED ACROSS YEARS ****

CLASS	AGE GROUP	NO.	PREGNANT NO.	%	LACTATING NO.	%
BRD COWS	19-21 MN	3	1	0.0	0	0.0
BRD COWS	22-24 MN	94	75	0.0	0	0.0
BRD COWS	31-36 MN	93	82	0.0	0	0.0
BRD COWS	4 YR	88	64	0.0	0	0.0
BRD COWS	5 YR	84	64	0.0	0	0.0
BRD COWS	6 YR	81	62	0.0	0	0.0
BRD COWS	7 YR	79	62	0.0	0	0.0
BRD COWS	8 YR	77	60	0.0	0	0.0
BRD COWS	9 YR	73	56	0.0	0	0.0
BRD COWS	10 YR	71	54	0.0	0	0.0
BRD COWS	TOTAL	741	580	0.0	0	0.0
REPL HFR	9 MN	3	0	0.0	0	0.0
REPL HFR	10 MN	21	0	0.0	0	0.0
REPL HFR	11 MN	31	0	0.0	0	0.0
REPL HFR	12 MN	45	0	0.0	0	0.0
REPL HFR	TOTAL	99	0	0.0	0	0.0

Figure 14. Average of Inventory Reports over the last ten years simulated.

The two final reports (Figure 13 and 14) contain information related to forage consumption and inventory (respectively). Entries in the feed consumption report are determined on a monthly basis for each herd, class and pasture. The inventory report is obtained at the first of the first month of the production year (the month calving is initiated).

There is flexibility in the number of reports produced and the amount of information contained in each of them for any given run. Although most of the reports that have been presented involved data generated over the last ten years of a simulation, the time period to be summarized, the frequency of generation of the reports and the number of reports are left almost totally to the discretion of the user. In addition to these summaries, four data files are created during each run: wean, feed, sales and inventory. These files are available for any further analyses that may be desired.

PHYSICAL REQUIREMENTS OF THE PROGRAM

Although our shift to the minicomputer environment has prompted a good deal of optimization in programming, the model and the associated output retains non-trivial space requirements. In order to obtain the amount of information that we consider minimal for interpretation and reasonable execution time, 2.2 megabytes of disk storage and .5 megabyte of available main memory are currently required.

VALIDATION

Since the TAMU model is designed to be adapted to varying sets of conditions, the procedure we follow is to validate for each new location. The validation process includes examination of the effects of the specific input parameters for the particular forage/ feed resource and cattle. Generally, when there is a close agreement (within 95% confidence intervals) between the simulated results and the actual data, the model is assumed validated and the effects of alternative inputs may then be examined with the model. However, if there is not close agreement, reassessment of the assumptions of the model is required and changes in the models are made so that the model more accurately reflects the system which it is intended to represent.

An example of the validation process was given by Kothmann and Smith (1983) for a cooperative Coastal Prairie ranch in Texas. Animal performance data from the cooperating ranch were used to validate the input parameters and that model structure closely simulated existing conditions. After the validation run was accepted, it was set as the baseline for comparison of management alternatives.

For this example, breedtype was specified in terms of genetic potential for mature size, milk production, and maturing rate of Santa Gertrudis cows. Mature weight (WMA) was set at 500 kg based on individual weights, and condition scores taken on 200 cows in mid-April and again in mid-June, 1979. The mean weights and condition scores (scale 1 to 9) were 372 kg and 3.76 units in April and 439 kg and 4.78 units in June. The relationship between weight and condition for each month was similar. WMA represented the weight of mature cows in good condition; hence, the weight for a condition score of about 7 was selected. Milk production potential (PMA) was set at 11.5 kg/day (peak day lactation potential at maturity).

In addition to the major specifications associated with breedtype, three other parameters were altered from those normally used for British breedtypes to better reflect the Zebu component. The percentage of mature size at which puberty could first occur was increased from 40 to 50 percent, and the upper limit for this percentage to affect puberty was increased from 60 to 70 percent. Also, preweaning and postweaning maturing rates were reduced.

As an example of forage input values, digestibilities of the base forage intake of cows, along with a seasonally "improved" set, for each month of the year are graphed in Figure 15. The correspondence between actual and simulated seasonal weight changes of mature cows is shown in Figure 16. The seasonal patterns of change for simulated and actual cow weights corresponded closely, but the magnitude of annual weight fluctuation was less for simulated data. The greater changes for actual weight were attributed to differences in gut fill not accounted for by the model assumption of constant fill and were similar in magnitude to the 67 kg difference observed between April and June for cows in similar condition. The cows were very drawn in early spring after the low level of winter hay feeding had been discontinued and before adequate spring growth of grass was available. On the other hand, the high availability of low quality forage resulted in near-capacity rumen fill in the fall. In addition, lean tissue loss resulting from protein deficiency was not accounted for in the model and may further explain why actual weights were less than simulated weights in the spring. The difference in magnitude of fluctuation between actual and simulated cows weights was not considered an important limitation to this study.

Simulated WM and W for replacement females from weaning through maturity along with a number of actual

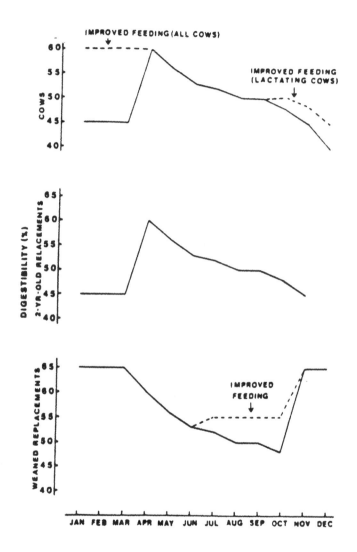

Figure 15. Organic matter digestibility of forage by month and class of stock for base (———) and improved (----) nutritional treatments.

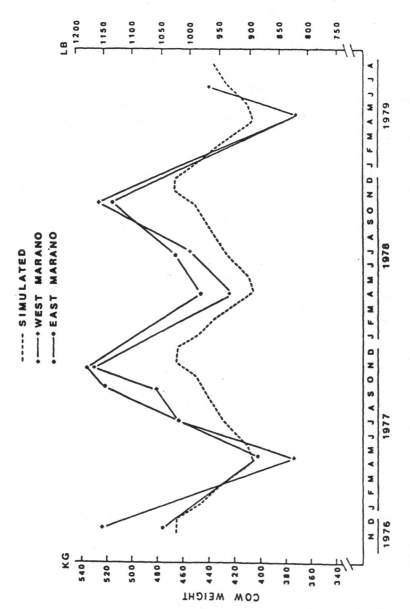

Figure 16. Actual and simulated seasonal changes in mature cow weights.

weights are shown in Figure 17. The actual weights were from a number of different pastures and years; hence, the lack of exact correspondence was not surprising. Except for the reduced magnitude of animal fluctuation discussed above, the correspondence between the actual and simulated data was considered good.

Comparisons of actual and simulated weights of nursing calves are shown in Figure 18 for birth months combined. Except for some yearly variation, the correspondence of weights near weaning was good. The model overestimated early spring weights, especially for calves born early in the calving season. This overestimation was considerably larger until the model was modified to increase maintenance requirements in young calves due to cold stress. The final simulation of preweaning weights was considered quite adequate.

Simulated postweaning growth rate of yearling heifers was very similar to actual growth rate (Table 1). The heavier actual weights in this table were attributable to the 30 to 40 percent culling of light heifers; whereas, the simulated weights represented an average of all heifers.

Calf death rates are summarized by birth month in Table 2. The much higher death rates of December- and January-born calves reflected the stress of cold, wet winters and inadequate nutrition for lactating cows. The correspondence between actual and simulated death rates by birth month was excellent. Cow death rates were also variable, but the simulated annual rate of 3.3 percent was identical to the 1978 ranch average of 3.3 percent.

Simulated pregnancy percentage was somewhat higher than the 1978 palpation records, averaged over all cow ages (Table 3). This difference was attributed to the incidence of brucellosis in the herd that was not accounted for by the model.

The high degree of correspondence between the baseline simulation and actual performance data was encouraging and suggested that useful comparisons of management alternatives could be made. In this particular example, after validation, eight simulations of a 4 x 2 factorial design with 4 livestock management schemes and 2 levels of nutrition were evaluated.

CONCLUSIONS

Jeffers (1978) pointed out that systems analysis is not a mathemtical technique, or even a set of mathematical techniques, but that it is instead a broad research strategy which seeks the solution of complex problems in a scientific and systematic manner. The system simulated by the Texas A&M Beef Cattle Simulation Model involves only the growth, reproduction and

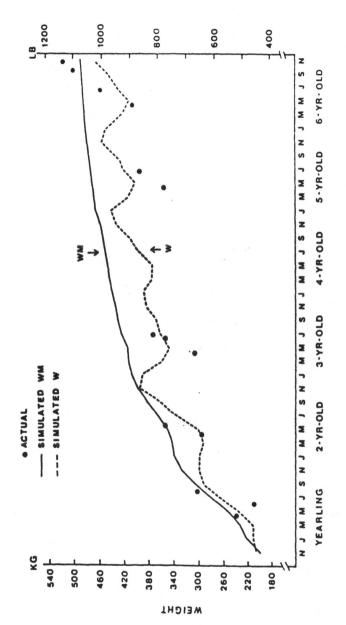

Figure 17. Simulated structural size (WM), weight (W) growth curves and actual weights at different ages.

186

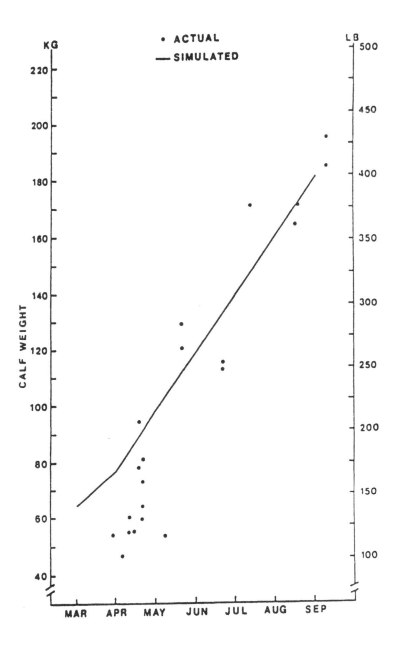

Figure 18. Actual and simulated preweaning calf
weights averaged across months.

Table 1.   Actual Weight (kg) and Gain (kg/day) of
            1977-born Yearling Heifers and Baseline
            Simulation.

Date	Actual		Date	Simulated	
	Weight	ADG		Weight	ADG
4/4/78	240		4/1/78	215	
		.55			.54
7/29/78	303		8/1/78	279	

Table 2.   Percentage of Calves Born by Month and Death
            Rates by Month of Birth for Actual Data for
            the Years 1977-1978 and Baseline Simulation.

Source	Birth month	Born		Died	
		No.	%	No.	%
Actual	Dec.	337	24.5	62	18.4
	Jan.	489	35.6	85	17.4
	Feb.	236	17.2	21	8.9
	Mar.	313	22.7	20	6.4
		1375	100.0	188	13.7
Simulated	Dec.	199	44.2	28	14.1
	Jan.	107	23.8	20	18.7
	Feb.	79	17.6	9	11.4
	Mar.	65	14.4	2	3.1
		450	100.0	59	13.1

Table 3.   Actual Pregnancy Data Age Class and Baseline
            Simulation.

Herd	Age	Number	Pregnancy % 1978
E. Marano	7,8	187	63
W. Marano	6,7	191	58
E. Big Past.	6,7,8	229	52
W. Big Past.	5,6,7,8	225	82
E. Humble	6,7	217	51
W. Humble	4,5,6	223	82
E. Sq. Tank	3,4	201	42
W. Sq. Tank	3,4	216	60
Mean			61.3

Simulation	
3	85.3
4	60.8
5	64.5
6	62.4
7	69.2
8	63.8
9	65.3
10	62.0
Mean	67.0

lactation of beef cattle. This model will not simulate the cause of forcings on the system or the economic system which impacts beef production strategies in the United States. For this reason, an interdisciplinary approach is used to extend beyond the limits of the biological model.

The TAMU Beef Cattle Model has been used to simulate production under a wide range of conditions (Davis, et al., 1976; ILCA, 1978; Ordonez-Vela, 1978; Notter, et al., 1979a,b,c; Gomez, 1980; Sullivan, et al., 1981; Nelsen, et al., 1982; Boyd, 1983; Abassa, 1984, Doren, et al., 1985, Stokes, et al., 1985). In every case where unique forage parameters were used, interaction between forage scientists and our systems group was required before a validation was obtained. Arbitrary adjustments of input parameters to attain closer correspondence between an actual and a simulated data set serves no legitimate purpose. Nonetheless, forage parameters are almost always, at least partially, estimates based on related data and/or experience. In defense of this time consuming process, we can only state that it has always been necessary for our simulations and that if the adjustments were simply arbitrary to bring some aspect of the simulation output into line with the real data, it very likely would throw some other aspect farther out of line. Even though the potential for abuse of this practice is somewhat self-limiting, like all data created by scientists, the final criterion is the integrity of the scientist.

In most of the applications cited above, the output from the baseline and from altered practices were subjected to economic analysis. Again this is an interactive process between the simulator and the economist (the topic of the following chapter). Depending upon the individuals involved and the objectives of a systems analysis study, either the forage scientist, the animal scientist or the economist may take the lead role, but in any case, coordinated input from these three areas is recommended.

REFERENCES

Abassa, K.P. 1984. "Systems Approach to Gobra Zebu Production in Dahra, Senegal." Ph.D. dissertation. University of Florida, Gainesville.

Baker, J.F. 1982. "Evaluation of Genotype-environment Interaction in Beef Cattle Production Systems using a Computer Simulation Model." Ph.D. dissertation. Texas A&M University, College Station.

Boyd, M.E. 1983. "Simulation of Genetic Differences in Fertility in Two Different Environments." Ph.D. dissertation. Texas A&M University, College Station.

Cartwright, T.C. 1982. "Developing Animal Breeding Strategies to Increase Herd Efficiency and to Cope with Change." The Fifth A.S. Nivison Memorial Address. University of New England, Armidale, New South Wales.

Davis, J.M., T.C. Cartwright, and J.O. Sanders. 1976. "Alternative Beef Production Systems in Guyana." Journal of Animal Science. 43:235 (Abstract).

Doren, P.E., C.R. Shumway, M.M. Kothmann, and T.C. Cartwright. 1985. "An Economic Evaluation of Simulated Biological Production of Beef Cattle." Journal of Animal Science. 60:913.

Forrester, J.W. 1968. Principles of Systems (2nd Edition). MIT Press, Cambridge, MA.

Gomez, F.G. 1980. "A System Analysis of Dual Dairy and Beef Production Herds." Ph.D. disseration. Texas A&M University, College Station.

ICLA. 1978. Mathematical Modelling of Livestock Production Systems: Application of the Texas A&M University Beef Cattle Production Model to Botswana. T.C. Cartwright, (Ed.). ICLA, Addis Ababa, Ethiopia.

Jeffers, J.N.R. 1978. An Introduction to Systems Analysis: With Ecological Applications. Edward Arnold Ltd., London.

Kothmann, M.M., and G.M. Smith. 1983. "Evaluating Management Alternatives with a Beef Production Systems Model." J. Range Management. 36:733.

Nelsen, T.C., T.C. Cartwright, A.K. Angirasa, and F.M. Rouquette, Jr. 1982. "Simulated Effect of Calving Season and Winter Hay Feeding Level on Cow Herd Productivity." Journal of Animal Science. 54:29.

Notter, D.R., J.O. Sanders, G.E. Dickerson, G.M. Smith, and T.C. Cartwright. 1979a. "Simulated Efficiency of Beef Production for a Midwestern Cow-calf Feedlot Management System. I. Milk Production." Journal of Animal Science. 49:70.

_____. 1979b. "Simulated Efficiency of Beef Production for a Midwestern Cow-calf Feedlot Management System. II. Mature Body Size." Journal of Animal Science. 49:83.

_____. 1979c. "Simulated Efficiency of Beef Production for a Midwestern Cow-calf Feedlot Management System. III. Crossbreeding Systems." Journal of Animal Science. 49:92.

Ordonez-Vela, J. 1978. "Systems Analysis of Beef Production in the Western High Plains of Venezuela." Ph.D. disseration. Texas A&M University, College Station.

Sanders, J.O., and T.C. Cartwright. 1979a. "A General Cattle Production Systems Model. I. Structure of the Model." Agricultural Systems. 4:217.

_____. 1979b. "A General Cattle Production Systems Model. II. Procedures Used for Simulating Animal Performance." Agricultural Systems. 4:289.

SAS Institute, Inc. 1982. SAS User's Guide: Basics, 1982 Edition. SAS Institute, Inc., Cary, N.C.

Stokes, K.W., C.R. Shumway, P.E. Doren, T.C. Nelsen, and T.C. Cartwright. 1986. "Economic Effects of Alternative Beef Cow Size and Milking Potential in Cow-calf Operations." Journal of Animal Science. 60:(in press).

Sullivan, G.M., T. C. Cartwright, and D.E. Farris. 1981. "Simulation of Production Systems in East Africa by Use of Interfaced Forage and Cattle Models." Agricultural Systems. 7:245.

VIII

Economic Analysis of Cattle Systems Using the Texas A&M Beef Cattle Simulation Model

Gregory Sullivan and Enrique Cappella

INTRODUCTION

In real-time, decision-making applications, the livestock manager requires a computational model which will predict system responses to an array of possible actions. The manager then weighs the cost of each alternative against anticipated returns. Because most biological systems are time variant, the range of system responses will vary. Complicating the picture further is the fact that plant growth and animal performance are highly dependent upon weather conditions which generally cannot be anticipated.

The Texas A&M University Beef Cattle Simulation (TAMU MODEL) has been widely used under various environmental conditions for simulating beef production. The biological functions have been detailed by Sanders and Cartwright (1979a, b) and in this volume by Cartwright and Doren (1986). The objective of this paper is to describe the use of the TAMU MODEL for economic analysis of production and marketing decisions.

This paper describes previous research using the TAMU MODEL for economic analysis. Review of previous work illustrates the applications for herd management. An important consideration is to examine the linkages between the TAMU MODEL and quantitative models commonly used by economists.

Advanced Marketing Systems, Auburn, Alabama and School of Veterinary Medicine, Auburn University, Alabama.

REVIEW OF SOME ECONOMIC APPLICATIONS OF THE TAMU MODEL

Some of the major contributions of the TAMU MODEL have been to analyze the micro-level decision-making process for livestock producers. These applications have been both partial enterprise analyses as well as whole farm planning models. Applications have been to production environments in both the United States and abroad. Only adaptations of the TAMU MODEL where economic analysis was the primary focus are presented.

ECONOMICS OF ALTERNATIVE BEEF CATTLE GENOTYPES

Research conducted by Stokes used the TAMU MODEL to examine the economics of selecting beef cattle genotypes (Stokes, 1980; Stokes et al., 1981). Key biological inputs to the TAMU MODEL were varied for different WMA and different levels of milk production: small (450 kg), medium (500 kg), and large (550 kg) frame cattle with three levels of milk production (light 8 kg/day, medium 11 kg/day, and heavy 14 kg/day). The TAMU MODEL was used to simulate preweaning and post-weaning performance of the nine different beef cattle genotypes. Prices for the classes were tied directly to sale month, sex, weight, condition (fatness), and year in the cattle cycle. The simulation model is limited in its capability to vary quality grades of classes so all calves were assumed to be choice grade.

Stokes estimated a function for price differences based on the condition of calves based on a ratio of weight (W) to skeletal frame size (WM). Thin calves received a premium if the index was .7 kg/framesize unit or less; the price received was increased one standard deviation above the mean. Price was reduced if the index was greater than 1.3 units indicating fleshly cattle; price was reduced one standard deviation. No correction was made when the index was equal to 1.0, and a linear relationship between fat level and price was assumed when index was between .7 and 1.3.

Four procedures were used by Stokes to project prices for evaluating the alternative production system: hindsight, current prices, USDA forecast of prices, and hedging using the futures market. Price data were collected during the period of 1972-1979 to test all procedures.

The research by Stokes found that selling weaned calves directly to the feedlot had a higher average net return per head ($38.67) compared to selling calves at weaning. The second largest average added net return above selling at weaning was to move weaned calves to winter grazing on wheat pasture followed by feedlot finishing ($16.61) (Farris et al., 1980). A herd of large frame-medium milk capacity cows was the preferred frame size for calves sold directly to the feedlot, but for winter grazing and follow-on feeding, a herd of large frame-heavy milkers was preferred ($26.36). Cost of production was estimated for the nine class sizes and six marketing strategies. The lowest cost of production was $67.79 per cwt for large-frame cattle with heavy milking capacity, and the highest cost was for a small-frame herd with light-milking capacity at $73.71 (Stokes et al., 1981).

Analysis of the performance for price forecasting methods found the USDA price forecast to be the most effective decision-making tool for profit determination. Hedging strategies incorporated into the model proved less effective than USDA price forecast for increasing average net returns.

ECONOMICS OF VERTICAL PRODUCTION SYSTEMS INCORPORATING RISK

The TAMU MODEL was used to analyze forage-beef production systems incorporating risk analysis for a representative farm in East Texas (Angirasa, 1979; Angirasa et al., 1981). The TAMU MODEL generated input-output coefficients for three forage alternatives: Coastal bermudagrass, Coastal bermudagrass overseeded with rye-ryegrass, and common bermudagrass overseeded with crimson clover-ryegrass. Several marketing plans of selling weaned calves, feeder calves, and grain-finished slaughter cattle were analyzed. Two alternative calving seasons (spring and fall) and four levels of winter feed availability (continuously available to heavy stress) were considered.

The TAMU MODEL generated forage consumption and animal performance data for each of the animal production systems. The TAMU MODEL focused on animal performance given a nutritional regime based on daily availability and quality of the forage.

Since different combinations of feed can be produced to meet the same nutritional level, a linear programming (LP) model was developed that analyzed alternative forage and supplemental feed systems for the simulated beef enterprise options. A given herd size for the representative farm was selected from the LP model based on the least cost combination of a single forage and supplemental feeds at alternative grain prices (Angirasa et al., 1981, p. 90). Cost of production for each pasture system was obtained from Texas Agricultural Extension Service crop budgets and was adjusted for management (Angirasa, pp. 59-61).

The LP model allowed for permutations in stress levels of cattle and prices of grain sorghum for feedlot finishing. Three LP models were utilized: a long run model subject only to a land constraint, a short-run LP model with constraints on resource vectors and treatment of fixed costs, and a MOTAD model. For the MOTAD model, risk was incorporated by including deviation in forage yield, beef price, and supplemental feed price variability. An E-A (return/absolute deviation) efficient set is derived by parameterizing the restraint on total absolute deviations in net returns (Hazell, 1971).

Results of the analysis showed that incremental changes in feeder calf or slaughter cattle prices improved the profitability of the stocker or finishing activities. The supply response to increased prices based on 1977 prices affected the beef activities. If prices fell below the 1977 "normal" prices, no production would take place. At 20% above 1977 "normal" prices the cow-calf operation with fall calving shifted to cow-stocker with spring calving. When prices were increased by 70 percent, a cow-drylot-finished program entered the solution (Angirasa, et al., 1981, p. 94).

The use of the LP model traced changes in activity mix, forage system, and beef supply response. The combination of the two models enhances the types of economic analysis necessary for forage/beef systems research. The use of LP models extends the TAMU MODEL and allows optimization of resource use.

ECONOMICS OF IMPROVED MANAGEMENT PRACTICES - WHOLE HERD SIMULATIONS

The TAMU MODEL was used to analyze traditional management systems and measure the impact of alternative production and marketing systems. Two examples are presented from research applying the model to production systems in Tanzania and Costa Rica.

Alternative Management Systems - Tanzania

Livestock production systems for two regions in Tanzania were examined using the TAMU MODEL (Sullivan, 1979; Sullivan et al., 1981; Sullivan et al., 1985). A representative village herd was estimated for two production zones: Sukumaland, south of Lake Victoria with an average rainfall of 800 mm per year and Gogoland in the central region of the country with a drier climate than Sukumaland with approximately 500mm of rainfall per year. The predominate animal breed is the Tanzania Shorthorn Zebu (TSZ), a hardy but small frame animal. The grass types in both regions were different with Themeda triandra, the predominate range grass in Sukumaland, and Panicum maximum in Gogoland.

The TAMU MODEL was interfaced with a forage model conceptualized and designed from a model by Smith and

Williams (1973). The forage model increased the rich-
ness of the TAMU MODEL by allowing a feedback mechanism
between the beef and forage sub-models to allow for the
effects of stocking rate and grazing pressure variables
to influence the state forage variables in the next
time period (see Figure 1). Stocking rate is influenced by herd numbers and composition which affects
forage growth, grazing of forage, and quality of biomass. Grazing pressure is affected by quality of forage and the nutrition requirement of the herd. Grazing
pressure directly impacts on the defoliation rate of
standing green and dry grass (Sullivan, 1983).

Four improved management plans were evaluated by
comparing the changes in outputs to that of the baseline herd for each village. Identification of appropriate management practices drew from research conducted in Botswana using the TAMU MODEL (ILCA, 1978).
A baseline herd was generated and validated against
village-level field data collected in 1975-76 in each
production region and from secondary experimental data
from research stations in Tanzania.

The four management plans were a nine-month breeding program compared to the current twelve-month con-
tinual breeding season; calves weaned at eight months
instead of ten months; supplementation of hay to weaned
calves in the stress months of the dry season; and a
seasonal sales policy of removing unproductive cows
from the herd.

Climatic weather conditions for each region were
collected and a normal distribution for rainfall with
means and standard deviations for each month were estimated. A lognormal distribution with means and standard deviations for pan evaporation and radiation were
estimated. After validation of the baseline herd for
the representative village herd to actual estimates of
weights and performance in each region using average
weather conditions, ten year simulations of each village herd with stochastic weather conditions were generated. The impact of each management practice was
measured in terms of increased revenues to the village
and valuations of the outputs of milk and meat for home
consumption.

Seasonal sales policies of removing animals from
the herd during key periods of the year were examined.
Pricing policies by the government that fixed prices
based on weight were introduced to show what benefit
might accure if an increased rate of culling took
place.

Results of the analysis are shown in Table 1. The
greatest impact on the representative herd in Sukumaland was a combination of a sales policy, supplemental
feeding, and a restricted breeding program of removing
bulls for three months to prevent calves during the dry
stress periods. The nutritional and net monetary val-

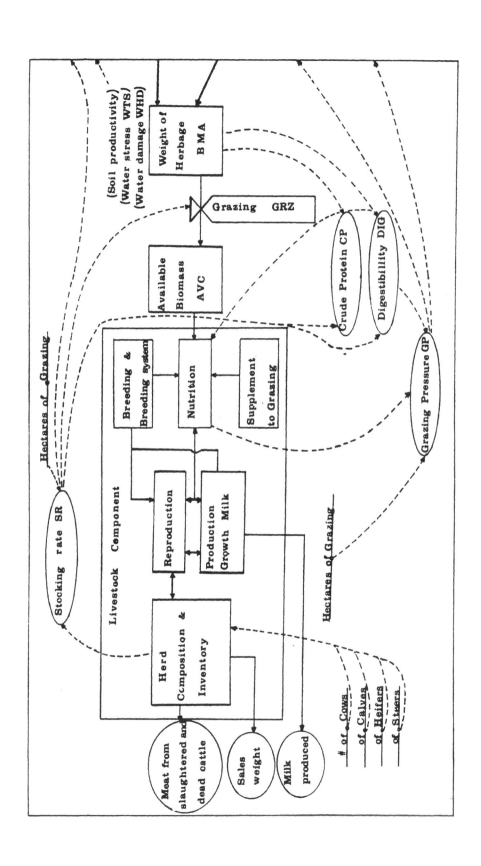

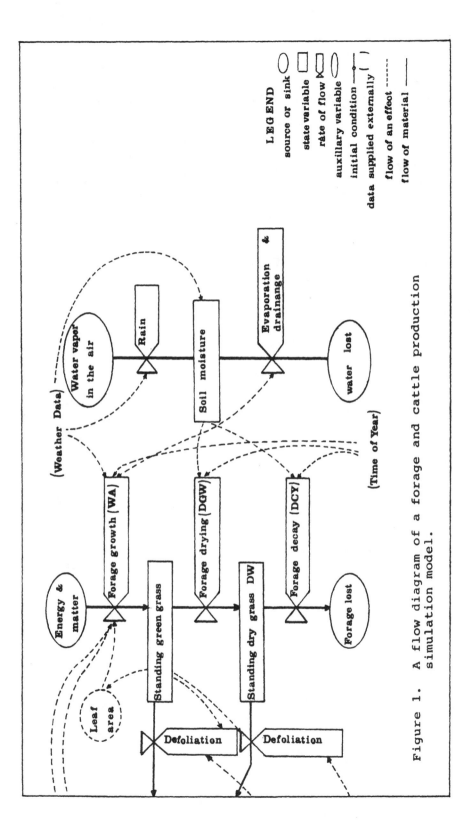

Figure 1. A flow diagram of a forage and cattle production simulation model.

LEGEND

⬯ source or sink
▭ state variable
◥ rate of flow
⬯ auxillary variable
◦─ initial condition
⎛ data supplied externally ⎞
---- flow of an effect
—— flow of material

Water vapor in the air

Rain

Soil moisture

Evaporation & drainange

water lost

(Weather Data)

Energy & matter

Forage growth (WA)

Standing green grass

Forage drying (DGW)

Standing dry grass DW

Forage decay (DCY)

Forage lost

(Time of Year)

Leaf area

Defoliation

Defoliation

Table 1. Average Annual Net Value of Output of Meat and Milk from Livestock Herds for Ten Year Simulation Period for Sukumaland and Gogoland in Tanzania.

Management Practice	Sukumaland			
	Calories/ Cap[a]	Protein Cap[a]	Annual Net Value	Percent of Baseline
	'000 Cal.	'000 grms.	'000 Tsh.	%
Baseline	47.72	3.42	780	100
Breeding season	46.29	3.32	760	97
Weaning at 8 months	47.72	3.42	781	100
Supplemental hay to calves	52.49	3.76	855	110
Weaning & breeding	46.77	3.35	764	98
Weaning & supplemental hay	46.29	3.32	759	97
Breeding & supplemental hay	51.06	3.66	835	107
Breeding & weaning & suppl.	45.01	3.28	748	96
Sales policy	56.31	4.04	918	118
Sales & supplemental & breeding[b]	60.13	4.31	984	126
Sales & supplemental hay[b]	---	---	---	---

Continued

Table 1. Average Annual Net Value of Output of Meat and Milk from Livestock Herds for Ten Year Simulation Period for Sukumaland and Gogoland in Tanzania (continued).

| Management Practice | Gogoland | | | |
	Calories/ Cap[a]	Protein Cap[a]	Annual Net Value[c]	Percent of Baseline
	'000 Cal.	'000 grms.	'000 Tsh.	%
Baseline	49.90	3.30	922	100
Breeding season	48.90	3.23	903	98
Weaning at 8 months	48.90	3.23	902	98
Supplemental hay to calves	75.85	5.02	1400	152
Weaning & breeding	47.90	3.17	883	96
Weaning & supplemental hay	45.91	3.04	844	92
Breeding & supplemental hay	74.35	4.92	1376	149
Breeding & weaning & suppl.	45.41	3.00	835	91
Sales policy	65.37	4.32	1205	131
Sales & supplemental & breeding[b]	---	---	---	---
Sales & supplemental hay[b]	63.37	4.19	1172	127

[a]The calculation of the nutritional value of milk produced assumes that the milking practice of dividing milk between the calf and home consumption remains constant. The value is the equivalent in calories and protein of milk used for home consumption. The nutritional value of meat is for animals consumed which were slaughtered or found dying in the herd.

[b]The sales polcy was considered with the combination of management practices that was felt to be relevant given their performance in each respective region.

[c]The value of meat and milk if animal products are sold in the village. The net value is adjusted for the cost of hay production for supplementing weaned calves where appropriate.

ues of meat and milk increased 26 percent above the baseline value. An important consideration is that the human population would not be made worse off because of a management practice. The combination of management practices enhanced the seasonal distribution of milk available from the herd for human consumption (Table 2). Percentage of milk production in the dry season increased from 9% to 23%.

Table 2. Seasonal Distribution of Milk for Village Consumption from the Cattle Herd for the 10-year Simulation Period.

	Wet	Season Dry	Total
	- - - - - (%) - - - - -		
Sukumaland			
Baseline	91	9	100
Sales, Supplies & Breed	77	23	100
Gogoland			
Baseline	56	44	100
Supplies Only	69	31	100
Supplies & Breed	58	42	100

The improved practices with the greatest impact in Gogoland were supplemental feeding to weaned calves with an increase in annual net value of 152 percent followed by a breeding program with supplemental feeding of 149 percent above the baseline results. The distribution of milk for home consumption was affected less by the combined breeding and supplemental feeding program (42%) compared to the baseline (44%) than for a supplemental program alone (31%).

The TAMU MODEL is capable of examining complex management strategies for a village herd as well as include important human requirements for meat and milk derived from the herd. The inclusion of a forage sub-model enriched the analysis for necessary herd control when forage is limiting.

Alternative Production Systems in Costa Rica

The TAMU MODEL was used to examine beef cattle systems in Costa Rica. Improved herd management can have an important impact on the industry, and the TAMU MODEL was used to examine alternative herd management strategies in the Dry Pacific Region of Costa Rica. The practices identified for analysis were: three (CNB), eight (SMB), and twelve month breeding managements (UNB).

Controlled breeding (CNB) was selected according to the breeding management done at the farm used for validation. Uncontrolled breeding (UNB) was selected according to traditional herd management for the country. The breeding strategy of restricting bulls from the cows for four months (SMB) (February - May) was established based on estimated monthly average weaning weights obtained with the uncontrolled breeding (Figure 2). Two weaning ages of six and eight months were selected for comparison of traditional and reduced weaning ages.

The major forage species in the Dry Pacific Region of Costa Rica is Hyparrhenia rufa. The nutritional values of the forage were available from secondary field data and were used to determine the monthly forage values required by the model. The cattle simulated were Brahman breed, a representative of the Bos indicus group.

The baseline herd (CNB2)[1] was simulated until herd composition, animal performance, and herd output was stable from year to year. This herd was validated against herd records collected from an agriculture experimental station located in the region. The same herd was used thereafter as the initial herd for the simulation of alternative management systems, which were run for 40 years.

Monthly market prices for the period 1975-1983 were averaged and used as index numbers to correct the average price per kilogram of cattle for 1983. The price was further adjusted for class and condition of the animal sold. The same technique used by Stokes (1980) and Sullivan (1979) using the ratio of weight (W) to the animal's genetic potential (WM) was used to adjust market price.

The characteristics of the herd dynamics throughout the 40 years of simulation for each breeding management are illustrated in Figures 3-5. The herd

---

[1]CNB1 denotes controlled breeding and weaning at six months, and CNB2 denotes controlled breeding and weaning at eight months.

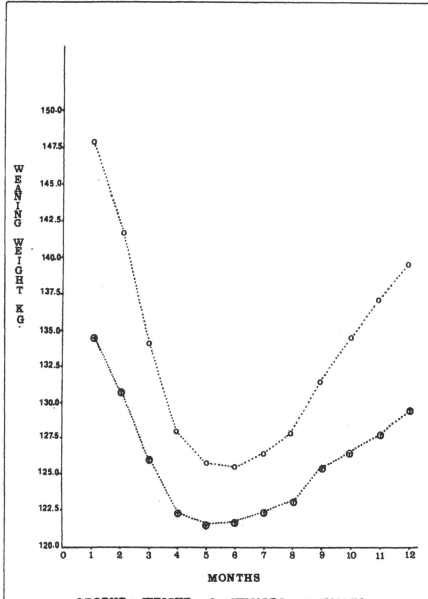

Figure 2. Simulated Monthly Weaning Weights for a
Representative Herd of the Dry Pacific
Region of Costa Rica with a Weaning Age
of Six Months and a Breeding Season of
Twelve Months/Year.

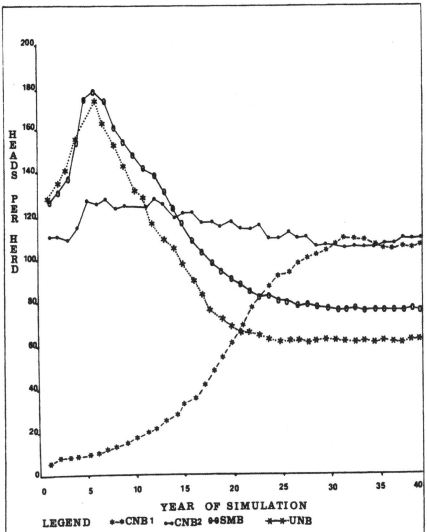

LEGEND *—*CNB1 •—•CNB2 ⊕⊕SMB *—*UNB

CNB1: Breeding is allowed during June, July and May.
      The 40th year corresponds to the initial herd
      for the simulation of CNB2, SMB and UNB.
CNB2: Breeding is allowed during June, July and May.
 SMB: Breeding is allowed during January, June, July,
      August, September, October, November, December.
 UNB: Breeding is allowed during twelve months/year.

Figure 3. Simulated Herd Sizes during 40 years for a
          Medium-Scale Herd under Controlled Breeding
          Management (CNB1), Overlayed with The Herd
          Sizes for Controlled (CNB2), Semicontrolled
          (SMB) and Uncontrolled (UNB) Breeding Manage-
          ments for the following 40 years.

growth follows a sigmoid curve leveling off at approximately 110 head by the thirtieth year. Herds with semi-controlled (SMB) and uncontrolled breeding (UNB) management at first increased and then steadily decreased to a lower herd size less than that under controlled breeding management (CNB2) (Figure 3).

The dynamics of calf mortality varied under different management plans (Figure 4). Higher calf mortalities occurred with uncontrolled breeding management (UNB) with a gradual increase to a peak of approximately 12 percent in the twelfth year and then a steady decline occurred leveling off at around 9 percent. The same pattern was visible for the semi-controlled breeding season (SMB) with a final lower mortality rate of seven percent after the 35th year.

The condition of the herd was tracked for forty years and measured by the ratio of the total weight of the animals in the herd (W) to their potential weight (W/WM) (Figure 5). Herds with a three month breeding management (CNB1 and CNB2) had a higher ratio (0.984) throughout the simulated periods. Herds with the lowest ratios (.957) were for semi-controlled (SMB) and uncontrolled breeding (UNB) programs.

Animal performance and herd output were examined at the fortieth year of simulation (Table 3). The average weight per head in the herd increased when a weaning age of six months was used. The shorter weaning age left mature cows in better condition, allowing them to begin cycling earlier and achieving better herd calving rates. The selling weights of steers and sale heifers were approximately the same. Greater revenues per animal unit resulted when weaning age was reduced.

The parameters considered also were affected by the length and time of the breeding season. The highest calving rate was observed for uncontrolled breeding management (UNB) followed by semicontrolled breeding management (SMB). Controlled breeding (CNB2) showed the lowest value (Table 3). The highest revenue per animal unit in the herd was realized from semi-controlled breeding management (SMB). Uncontrolled breeding (UNB) with a weaning age of six months generated only slightly less revenue per animal unit and the lowest value was generated for controlled breeding (CNB2).

The results found that the reduction of the herd size and the net revenue per animal unit of the simulated herds were controversial; an analysis of the present value of the future income stream was conducted. A discount rate (18%) was applied to the total revenue of the herd for the current year of simulation. The pattern of the present value of future net revenues (PV) when the weaning age was six months, was selected for the illustration of the events occurring during the 40 years of simulation (Figure 6).

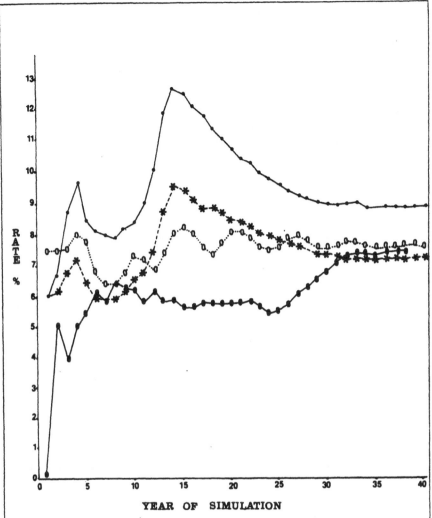

YEAR OF SIMULATION

LEGEND   ••••CNB1·   *-*CNB2   ··0·0·0SMB   •••• UNB

CNB1: Breeding is allowed during June, July and May. The 40th year corresponds to the Initial herd for the simulation of CNB2, SMB and UNB.

CNB2: Breeding is allowed during June, July and May.

SMB: Breeding is allowed during January, June, July, August, September, October, November, December.

UNB: Breeding is allowed during twelve months/year.

Figure 4. Simulated Calf Mortalities during 40 years for a Medium-Scale Herd under Controlled Breeding Management (CNB1), Overlayed with the Calf Mortalities for Controlled (CNB2), Semicontrolled (SMB) and Uncontrolled (UNB) Breeding Managements for the following 40 years.

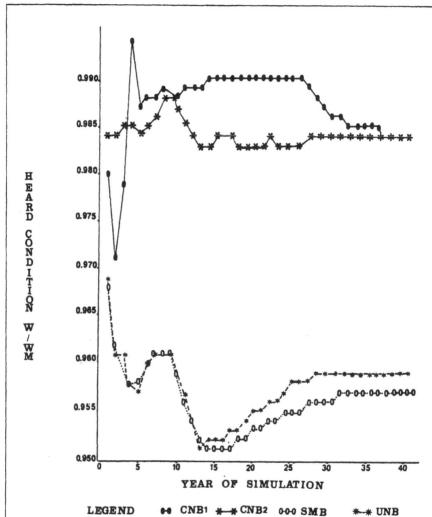

YEAR OF SIMULATION

LEGEND   ●–● CNB1   *–* CNB2   o-o-o SMB   *–* UNB

CNB1: Breeding is allowed during June, July and May. The 40th year corresponds to the initial herd for the simulation of CNB2, SMB and UNB.

CNB2: Breeding is allowed during June, July and May.

SMB: Breeding is allowed during January, June, July, August, September October, November, December.

UNB: Breeding is allowed during twelve months/year.

Figure 5. Simulated Herd Conditions during 40 years for a Medium-Scale Herd Under Controlled Breeding Management (CNB1) Overlayed with Herd Conditions for Controlled (CNB2), Semicontrolled (SMB) and Uncontrolled (UNB) Breeding Managements for the following 40 years.

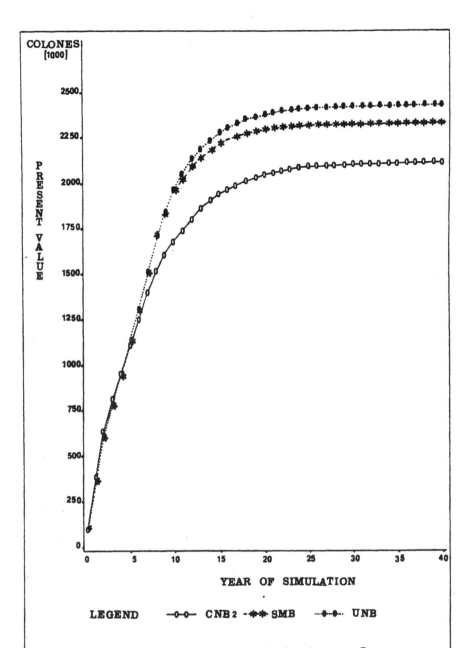

COLONES
[1000]

PRESENT VALUE

YEAR OF SIMULATION

LEGEND   —o-o— CNB2 -*-*- SMB   ···•-•··· UNB

Figure 6.   Present Value of the Future Revenue
            Stream for 40 years of Simulation for a
            Medium Size Herd Under Controlled Breed-
            ing (CNB) Semicontrolled Breeding (SMB)
            and Uncontrolled Breeding (UNB) Manage-
            ments and Weaning at Six Months of Age.

Table 3. Simulated Animal Performance and Herd Output in the 40th Year for Two Weaning Strategies and Three Breeding Management Alternatives for a Medium Scale Herd in the Dry Pacific Region of Costa Rica.

A

	Herd size Six[a]	Eight[a]	Weight/head Six	Eight	Mature cow Six	Eight
	(No.)		(kg)		(kg)	
CNB[b]	113	122	344	309	459	440
SMB[c]	81	92	307	269	444	430
UNB[d]	66	76	298	253	445	431

B

	Calving rate Six	Eight	Calf Mortality Six	Eight	Extraction Six	Eight
	(%)		(%)		(%)	
CNB	60.7	59.7	7.7	9.9	22.1	19.0
SMB	73.5	68.9	7.4	9.1	22.0	19.0
UNB	75.7	72.3	8.3	9.5	20.9	18.2

C

	Selling weight Six	Eight	Total Revenue Six	Eight	Revenue/AU Six	Eight
	(kg)		(x1000 colones)		(colones)	
CNB	424	421	353.5	333.1	3187	3098
SMB	422	417	251.5	230.8	3525	3256
UNB	422	419	194.1	173.5	3478	3168

[a]Weaning age in months

[b]CNB denotes breeding allowed during June, July, August.

[c]SMB denotes breeding allowed during February, March, April and May.

[d]UNB denotes breeding allowed year around.

It can be observed that the PV is similar for the three breeding management alternatives for the first six years. After this year, there is a constant tendency for a better PV when the SMB and UNB alternatives were used. After the 13th year, the highest PV is given by the SMB followed by the UNB. The CNB remained lower for this last period of observation (years 13 through 40). The present value of the future revenue stream for each management system in the 40th year is illustrated in the Table 4.

Table 4. Present Value of the Gross Revenue in the 40th Year of Simulation for a Representative Medium Scale Herd under the Environmental Conditions of the Dry Pacific Region of Costa Rica.

Breeding[b]	Present Value[a]	
	Six[c]	Eight[c]
	(x1000 colones)	
CNB	2109	1974
SMB	2428	2176
UNB	2329	2021

[a]Discount rate = 18%.

[b]CNB denotes controlled breeding management.
 SMB denotes semicontrolled breeding management.
 UNB denotes uncontrolled breeding management.

[c]Weaning age (months).

The results showed that the present value of the future revenue stream (PV) of the annual gross income is affected by weaning strategy and breeding management. The highest PV at the 40th year was obtained when weaning age was six months and decreased when it was eight months. According to breeding management, the best PV was obtained when semicontrolled breeding (SMB) was used, followed by uncontrolled breeding (UNB). The lowest PV was obtained with controlled breeding season (CNB). The use of a controlled breeding season assumes the possibility of the separation of the bulls during the desired months and the potential additional cost of this practice was not taken into account. If additional costs were to be applied, the difference in the benefits of the simulated herds will be larger.

In summary, the results indicated that:

(i) Weaning age and breeding management affect the animal performance of the simulated representative medium-scale herds under the conditions of the Dry Pacific Region of Costa Rica;

(ii) An earlier weaning age (six months) provided higher animal performance, herd output, and present value of the future revenue stream of the herds;

(iii) The present value of the future revenue stream was affected by breeding management. During the first six years of simulation, the present value was similar for the three breeding management alternatives tested. After the sixth year, the present value of semicontrolled and uncontrolled breeding managements remained similar but higher than the controlled breeding option. After the 13th year there was a tendency for higher present value for semicontrolled breeding management, followed by uncontrolled and controlled breeding managements.

The tendency indicated by the differences in the present value may be an indication that the establishment of short breeding seasons may have negative influences upon the profits of the beef cattle operations under the conditions of the DP of Costa Rica. It was concluded that the recommendations made by public and private institutions of the country, with regard to the management of the breeding time and length, as a means of improving cattle production in the DP, should be considered carefully, and that probably other management alternatives should be studied first.

ALABAMA RESEARCH USING THE TAMU MODEL -
STOCKER CATTLE MODEL

The TAMU MODEL was applied to research on forage/cattle systems in Alabama. An important cattle system developed in Alabama is growing out of stocker cattle on winter forages of small grains and legumes. This enterprise is important to the state because of the favorable climate for year-round production of forage and the availability of stocker calves. Vertical integration is an important enterprise for backgrounding and finishing on forage or with limited on-farm grain-feeding.

Research by Aderogba (1984) used the TAMU MODEL
interfaced with a linear programming (LP) model to
examine marketing strategies for winter grazed stocker
cattle. The model examined steers and heifers with
three starting weights (350, 450, and 550 pounds) grazing winter pastures of rye, rye grass and clover.
Cattle could be sold directly after grazing in either
March, April, or May or continued on limited-grain
feeding on the farm for 30, 60, 90, or 120 days. The
marketing decision was based on using the mean and
absolute deviation of prices for each weight class over
the period 1960 to 1983.

Information on animal's beginning and ending
weights with the average daily gains (ADG) were obtained from 10 years (1973-83) of experiments conducted at
the Auburn University Experiment Station at the Lower
Coastal Plains Substation (Harris et al., 1971, 1980).
Based on 10 years (1973-83) of animal performance data,
and four years of forage quality data, cumulative probability functions for forage availability in each month
were generated based on field data (Harris, unpublished). Forage availability was the random factor and was
chosen from a univariate density function for each
month of the year. A probability function was estimated from eight years of animal weights and results were
validated against two years of separate data. Variations in availability for the eight years were used as
the basis for the cumulative function. The estimated
ADG for the two test-years were not statistically different using the Smirnov statistic (Conover, 1980) from
the actual weights recorded (Aderogba, 1984). The
biological simulation model served as a means of incorporating stochastic occurrences to include risk in the
analysis. Performance and feed intake of steers and
heifers were predicted for the periods of November-
March 31, April 1-30, and for May 1-31 as a continuous
grazing operation. Following each grazing period, animal performance and feed intake were predicted for 30,
60, 90, or 120 days of feeding using the TAMU MODEL
(Table 5).

A linear programming model (MOTAD) was used to
evaluate strategies based on expected net returns (E),
and total absolute deviation of returns (A). Production activities included both animal and forage alternatives. Deviations in forage yield deviations estimated the risk during the production periods. At the
end of each grazing period, mean weights and year-to-
year deviations from the mean were calculated. The
deviations were a result of stochastic behavior of
forage availability. Annual yield deviations were
multiplied by the corresponding average selling price
for the period giving the dollar values of the deviations. The result was entered into the risk portion of
the model under the livestock grazing activities.

The estimated cost of producing rye-ryegrass-clover forage was $101.00 per acre. The average TDN requirements per animal per period on grazing and on feed following grazing were obtained from the biological simulation results. Animal performance data, including feed intake, were also predicted by using the TAMU MODEL. Average ending weights of the animals at the end of each period of grazing and feeding programs were used in the LP model (see Table 5).

Average monthly prices for the period 1961-83 from the Montgomery Auction Market were used to develop budgets for each production period and program, and for each animal category and sex (Aderogba, 1984). All prices were adjusted to be equivalent to 1983 values, using the USDA annual index of prices paid for factors of production (1977=100) (USDA). The adjusted cost per ton of formulated ration was estimated to be $143 per ton (Alabama Crop and Livestock Reporting Service). The cost deviation from the average for each year represented the risk for purchase of the ration. All other growing costs were for grazing or feedlot activities. Twenty-three years of deviations in buying or selling prices for livestock and feed expressed the risk for each respective activity in the LP model.

Results and Analysis

The model was validated and verified by running six separate LP models for each of the three weight classes and two sex categories and then a generalized model with all activities included. The results indicated consistency of the model's results. All categories of animals were kept on grazing until the end of May and followed by 30 days of feeding, except for medium and heavy steers and heifers, which were sold directly after May grazing. The amount of pastureland used per animal is also evidence of consistency. The heavy steers and heifers were stocked at 1.12 head per acre with light weight animals stocked at 1.40 head per acre which confirms that heavier animals lower stocking rate. The coefficient of variability (CV = A/E) for heavy, medium, and light weight steers were .95, .84, and .71, respectively. The CV's for heavy, medium, and light heifers were 1.72, 2.82, and 1.14, respectively. Steers had higher net returns and lower CV's than heifers. The overall lowest CV was for light weight steers of 71 cents for every dollar earned and had the greatest net return of $42,023 for the enterprise. Light weight heifers had a lower CV than for the heavier weights in that sex category.

Following the model's validation, six activities were combined into a single optimization program. Each livestock activity was able to compete for the availa-

Table 5. Animal Performances & Their Nutrient Requirements as Obtained From the Biological Simulation Model for Both Grazing and Feedlot Operations.

Animal Categories	Periods	Init. Wt. Grazing (LBS)	End. Wt. Grazing (LBS)	End. Wt. on Feed (LB) (Days)			
				30	60	90	120
Light Steers	Nov-March	350	610	679	744	810	875
	April	610	672	741	807	872	937
	May	672	733	802	867	933	998
Medium Steers	Nov-March	450	719	788	853	919	984
	April	719	782	851	917	982	1048
	May	782	838	907	937	1038	1103
Heavy Steers	Nov-March	550	816	885	991	1016	1081
	April	816	877	946	1012	1078	1148
	May	877	929	999	1065	1130	1198
Light Heifers	Nov-March	350	592	655	714	773	831
	April	592	651	713	772	830	888
	May	651	707	769	827	885	943
Medium Heifers	Nov-March	450	703	765	830	881	939
	April	703	760	822	880	937	995
	May	760	811	873	930	988	1045
Heavy Heifers	Nov-March	550	796	858	915	973	1030
	April	796	850	911	959	1026	1083
	May	850	899	960	1017	1075	1132

Continued

Table 5. Animal Performances & Their Nutrient Requirements as Obtained From the Biological Simulation Model for Both Grazing and Feedlot Operations (continued).

Animal Categories	Periods	Average TDN(LB) Grazing			Requirements/Head Feedlot (Days)			
		March	April	May	30	60	90	120
Light Steers	Nov-March	1170			360	730	1100	1500
	April		323		390	780	1200	1600
	May			324	420	837	1264	1700
Medium Steers	Nov-March	1324			400	800	1230	1600
	April		358		430	860	1300	1770
	May			340	460	900	1380	1800
Heavy Steers	Nov-March	1469			450	902	1363	1817
	April		384		478	950	1430	1920
	May			356	490	980	1470	1960
Light Heifers	Nov-March	1153			350	710	1080	1465
	April		314		370	750	1140	1540
	May			315	401	798	1200	1600
Medium Heifers	Nov-March	1336			390	780	1180	1560
	April		346		422	837	1262	1690
	May			334	440	870	1315	1760
Heavy Heifers	Nov-March	1458			430	860	1290	1790
	April		369		450	900	1350	1817
	May			348	474	935	1404	1880

ble resources under this condition. The generalized model's results, with unconstrained risk, indicated that the light weight steer option was the most profitable. The number of light weight steers produced and the expected net returns at the maximum risk level remained the same in the solution for the general model as was earlier presented in the validation results.

Within the steer category, lower risk and higher return associated with light weight steers could be attributed to their lower negative and higher positive price variations within the 23-year study period. The frequency and/or the magnitude of deviations within the study period's price data were influential factors in determining the probability levels at which an activity could enter the optimal solution. From this analysis, the light weight steers predominated over all other available activity choices.

Farm Plans and Possible Enterprise Combinations

A series of efficient farm plans were generated for the development of an E-A frontier. The net return (E) was parameterized between zero and infinity in increments of $2,100 to estimate the risk-return trade-offs and accompanied shifts in enterprise mix. The last farm plan (14th) generated a positive net return management. The choice of any of the farm plans estimated depends on the individual producer's preferred risk level.

Plan 1 had the highest net return of $42,023 and the highest amount of risk of $29,818 (Table 6). The number of light weight steers on grazing from November through May was 454. On June 1, the 454 feeders were placed in an on-farm feeding program and were subsequently sold at an average weight of 802 pounds at the end of June. All 325 acres of available pastureland were utilized and 136 tons of feed were purchased and used. The feed per gain was calculated to be approximately 8.7 lbs animal/lb feed.

The financial characteristics of the farm plans were considered by using both the coefficient of variabilities (CV's) and the Marginal Risk-Income Coefficient (MRIC), which is the inverse of the slope of the E-A frontier curve. The CV reveals the elements of risk involved per dollar earned, and MRIC ($\Delta A/\Delta E$) shows the increase in risk (A) for a unit increase in expected net return (E). Every dollar earned in plan 1 includes risk for producing valued at 71 cents. On the other hand, to move from plan 2 to 1, a one dollar increase in net returns involved $1.18 of risk.

MRIC allows for comparison to be made between two plans on how rapidly an increase or decrease in risk occurs from one critical point to another. For plans 4

Table 6. Return and Risk Tradeoffs for Farm Organizations With E-A Efficient Sets.

Farm Plans[a]	Previous November to	Beef Cattle Production and Marketing							Forage and Feed	
		No. of Head Grazed			No. of Head in Feedlot		Mean Net Returns	MRIC[c]	Rye-Ryegrass Clover (Acres)	Feedlot Ration (Tons)
		March	April	May	30[b]	90[b]				
1	March	454					42023.75	1.18	325.00	136.29
	April		454		454					
	May			454						
2	March	454					39923.70	1.13	325.00	150.20
	April		450		415					
	May			415		39				
3	March	437					37823.70	1.11	312.60	158.19
	April		437		388					
	May			388		49				
4	March	413					35723.70	.63	295.32	149.41
	April		413		367					
	May			367		46				
5	March	388					33623.70	.63	277.96	140.62
	April		388		345					
	May			345		43				
6	March	364					31523.70	.63	260.60	131.40
	April		354		324					
	May			324		40				
7	March	340					29423.70	.63	243.24	123.06
	April		340		302					
	May			302		28				

8	March	316							
	April	316	281		35	27323.70	.63	225.88	114.71
	May		281	281					
9	March	291							
	April	291	259		32	25223.70	.63	208.52	105.49
	May		259	239					
10	March	267							
	April	267	238		29	23123.70	.63	191.16	96.71
	May		238						
11	March	243							
	April	243	216		27	21023.70	.63	173.80	87.93
	May		216	216					
12	March	219							
	April	219	194		25	18923.70	.63	156.44	79.14
	May		194	194					
13	March	194							
	April	194	173		21	16823.70	.63	139.08	70.36
	May		173	173					
14	March	170							
	April	170	151		19	14723.70	.63	121.72	61.50
	May		151	151					

[a]Only light weight steers.

[b]Number of days in feedlot.

[c]Marginal risk income coefficient.

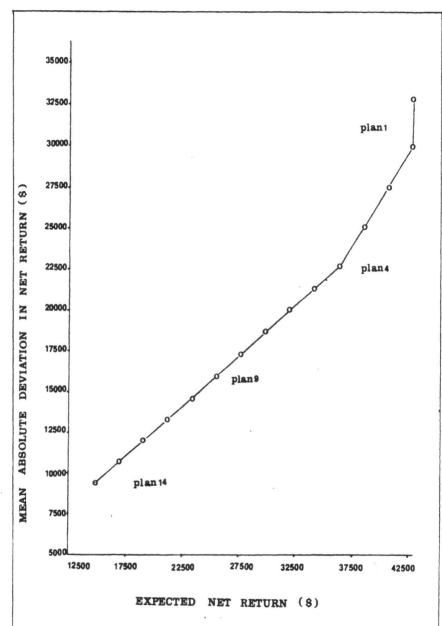

Figure 7. Efficient sets of E-A frontier for a
representative slaughter beef producer
in the Lower Coastal plains region of
Alabama.

to 14, the MRIC is a constant, meaning that for any amount of dollar increase in net returns, the dollar increase in risk remains the same. (See Figure 7 for relationship between net returns and risk.) The number of steers grazed and fed decreases along with the risk level. At the end of the April grazing period, 39 steers were placed on feed for 90 days in plan 2 and were subsequently sold in July, but in plan 2, like in plan 1, 454 steers were grazed from November through April. The 90 days feeding operation continues through all farm plans with different number of steers in solution.

STRENGTHS AND WEAKNESS OF TAMU MODEL
FOR ECONOMIC ANALYSIS

The TAMU MODEL has proven to ba a valuable tool for economic analysis of production impacts on livestock herds. The use of the biological model for benefit-cost analysis of improvements can be made to evaluate alternatives that otherwise would be too costly to test. The TAMU MODEL can be the foundation for adding peripheral programs to enhance economic analysis to include a more comprehensive study of the problem.

Some TAMU MODEL enhancement procedures can incorporate risk analysis from climatic conditions. Incorporation of risk from weather can be done in several ways. First, the input values of crude problem (CP), digestibility (DIG), and availability (AVC) can be adjusted to reflect environmental changes on the herd. These values can be held constant for each month throughout the simulation or can change exogenously to the model. This procedure of altering forage characteristics is used to examine periods of varying forage conditions.

Climatic conditions can also be characterized as stochastic variables to allow for random influences on the herd's performance. This effect is important when considering the risk associated with changing from current practices to new practices. The possibility of failure is an important consideration why producers do not want to change their practices. Stochastic climatic variables can affect the quality (CP or DIG) or availability (AVC) variables (Aderogba, 1984). The influence of climatic variables can be filtered through soil moisture and temperature response functions in a forage submodel (Sullivan, 1983; Sullivan et al., 1981b). The effects of variability in weather can be measured by economic valuation of the output of the simulation model.

The TAMU MODEL can be enhanced to examine livestock management through an interfaced forage model. Economic decisions concerning herd size, stocking rates, put and take systems based on height, or standing biomass of the forage (Sullivan and Huck, 1985) can improve the decision-making capabilities of the model. In the Southeastern United States, improved forage pastures are costly to establish and timely application of fertilizer is critical. The TAMU MODEL can be useful for economic analysis of these decisions.

The interface of a linear programming model with the TAMU MODEL provides a broad use for whole farm planning to include the competition between crops, forage, and livestock activities for available resources. Economists have quantitative tools that are useful in collaborating with the animal scientists. Issues of resource allocation critical to profitability of the farm can be examined. This is especially true for farm growth analysis since the TAMU MODEL is time variant and economic analysis of farm growth and investment decisions are important. The extension of a linear programming model or a farm growth simulation model allow for optimization, whereas the TAMU MODEL by itself cannot be optimized because of the many variables influencing the herd.

A general weakness of a biological model like the TAMU MODEL is exclusion of key variables regarding if particular management practices are capable of being adopted. Concern is raised about the suitability of recommendation to be implemented. Obviously, the TAMU MODEL is a means to an end - better managed and productive herds - yet producers may continue to adhere to their traditional practices. This concern is especially true for livestock production under extensive grazing systems. Detailed economic data on all costs, especially the manager's time, need full accounting before recommending certain practices. Adjusting the breeding program from a continuous to a restricted breeding season incurs extra costs in labor and facilities and it assumes available fenced pastures. All these costs need to be accounted. The economist working with the animal scientist can ensure accurate appraisal of any recommended practice.

The outlook for increased applications and refinement of the TAMU MODEL is promising. The extension of the MODEL from simulating classes of animals to tracking the performance of individual animals (Baker, 1982) will continue to establish the MODEL as a important tool for beef cattle systems research (Kothman and Smith, 1983). The TAMU MODEL will be a catalyst for attracting scientists with an interdisciplinary orientation to collaborate on systems research.

REFERENCES

Aderogba, Kolajo A. 1984. "Evaluation of Market Strat-
    egies Under Risk for Beef Produced on Winter Forage
    Systems in the Lower Coastal Plains of Alabama."
    Unpublished M.S. thesis, Auburn University, Ala-
    bama.

Aderogba, K.A., G.M. Sullivan, R.R. Harris, and N.R.
    Martin, Jr. 1985. "Calf Marketing Strategies for
    Livestock Producers in Alabama: Interfacing a
    Bioeconomic and a Linear Programming Model." Se-
    lected paper presented at the AAEA Annual Meetings,
    Iowa State University, Ames Iowa.

Alabama Crop and Livestock Reporting Service. "Alabama
    Agricultural Statistics". Montgomery, Alabama,
    selected issues.

Angirasa, Aditi K., C. Richard Shumway, T.C. Nelson,
    and T.C. Cartwright. 1981. "Integration, Risk, and
    Supply Response: Simulation and Linear Programming
    Analysis of an East Texas Cow-Calf Producer." So.
    J. of Agr. Econ.. 13:89-98.

Angirasa, A.K. 1979. "Firm Level Beef Supply: A Simu-
    lation and Linear Programming Application in East
    Texas." Ph.D. disseratation, Texas A & M Univer-
    sity, College Station.

Anthony, W.B., J.G. Starling, C.A. Brogden, E.L. May-
    ton, and H.E. Burgess. 1970. "Slaughter Steers
    from Winter Grazing with Supplemental Feeding."
    Highlights of Agricultural Research, Vol. 17, Agri-
    cultural Experiment Station, of Auburn University,
    Auburn, Alabama.

Baker, J.F. 1982. "Evaluation of Genotype-Environment
    Interactions in Beef Cattle Production Systems
    Using a Computer Simulation Model." Ph.D. thesis,
    Texas A&M University, College Station.

Cartwright, T.C., and P.E. Doren. 1986. "The Texas A&M
    Beef Cattle Simulation Model." in Simulation of
    Beef Cattle Production Systems and Its Use in Eco-
    nomic Analysis. T.H. Spreen and D. Laughlin
    (Eds.). Westview Press, Boulder.

Conover, W.J. 1980. Practical Nonparametric Statis-
    tics, 2nd ed. John Wiley & Sons, New York.

Farris, D.E., K.W. Stokes, T.C. Cartwright, and T.C.
Nelson. 1981. "Own 'Em All the Way." Beef. p.
48.

Harris, R.R., W.B. Anthony, V.L. Brown, J.K. Boseck,
H.F. Yates, W.B. Webster, and J.E. Barrett, Jr.
1971. "Cool-Season Annual Grazing Crops for Stocker Calves." Bulletin No. 476. Agricultural Experiment Station, Auburn University, Auburn, Alabama.

Harris, R.R., J.A. Little, V.L. Brown, and G.M. Sullivan. 1980. "Finishing Steers on Cool Season Grazing." Highlights of Agric. Research, Vol. 27.
Agricultural Experiment Station, Auburn University,
Auburn, Alabama.

Hazell, P.B.R. 1971. "A Linear Alternative to Quad-
ratic and Semivariance Programming for Farming
Planning Under Uncertainty." Amer. J. of Agr.
Econ. 53: 53-62.

International Livestock Center for Africa (ILCA).
Mathematical Modeling of Livestock Production Sys-
tems: Application of Texas A&M University Beef
Cattle Production Model to Botswana. ILCA System
Study No. 1, Addis Ababa.

Kothman, K.K., and Gerald M. Smith. 1983. "Evaluation
Management Alternatives with a Beef Production
Systems Model." J. of Range Management. 36:733-
740.

Kropf, W., S. Aragon, N. Kunzi, and W. Hagnauer. 1983.
"Dairy Ranching in Costa Rica." World Animal Review. January-March, pp. 23-27.

Sanders, J.O., and T.C. Cartwright. 1979a. "A General
Cattle Production Systems Model I: Structure of the
Model." Agricultural Systems. 4:217-27.

_____. 1979b. "A General Cattle Production
System Model II: Procedure Used for Simulating
Animal Performance." Agricultural Systems. 4:289-
309.

Smith, K.C.G., and W.A. Williams. 1973. "Model Development for a Deferred-Grazing System." J. of Range
Management. 26:454-460.

Stokes, K.W., D.E. Farris, and T.C.Cartwright. 1981.
"Economics of Alternative Beef Cattle Genotype and
Management/Marketing Systems." So. J. of Agr.
Econ. 13:1-10.

Stokes, K.W. 1980. "Economics of Alternative Beef Genotypes and Cattle Management/Marketing Systems." Ph.D. disseratation, Texas A & M University, College Station.

Sullivan, G.M. 1979. "Economics of Improved Management for Transforming the Forage/Livestock Systems in Tanzania - A Simulation Model." Ph.D. dissertation, Texas A&M University. College Station.

_____. 1983. "Modeling Livestock Management Systems Under Variable Climatic Conditions in East Africa." in Analysis of Biological Systems: State-of-the-Art in Ecological Modeling. Eds. W.K. Lauenroth, G.V. Skogerboe, and M. Flug, pp. 411-417. Elsenier Press.

Sullivan, G.M., T.C. Cartwright, and D.E. Farris. 1981a. "Simulation of Production Systems in East Africa by Use of Interfaced Forage and Cattle Models." Agricultural Systems. 7:245-265.

Sullivan, G.M., M.C. Huck, C.S. Hoveland, and R.L. Haaland. 1981b "An Interactive Pasture/Cattle Production System Simulation Model." Abstract in XIV International Grassland Congress Proceedings, Lexington, Kentucky.

Sullivan, G.M. and M.G. Huck. 1985. "Modeling a Forage/Beef System for Management Optimization." unpublished manuscript.

Sullivan, G.M., D.E. Farris, and J.R. Simpson. 1985. "Production Effects of Improved Management Practices in East African Cattle Grazing Systems." Quarterly Journal of International Agriculture. 24:22-37.

United States Department of Agriculture. "Alabama Prices (Annual Summary)." Statistical Reporting Service, USDA. Washington, D.C. selected issues.

IX

The Colorado State University Beef Production Model

R.M. Bourdon and J.S. Brinks

HISTORICAL PERSPECTIVE

The CSU model is not original. It is a customized version of the Texas A&M class model developed by Sanders and Cartwright (1979a,b) and described in an earlier paper. Being animal breeders, we originally intended to use the model to evaluate genotypes and generate economic weights for a number of traits. Results of this kind cannot exist independent of management, however, so we have necessarily used the model to evaluate management alternatives as well.

The CSU model has been evaluated for "reasonableness", but has not been rigorously validated against a specific set of animal and forage data, as has the TAMU model. Consequently, the model is intended to show trends and relationships and not to provide precise answers in particular situations. Future versions of the model should have this capability.

CHANGES IN THE BIOLOGICAL MODEL

The CSU version of the TAMU model differs from the original in a number of respects. It incorporates most of the modifications added by Notter, (1977) particularly those relating to feed intake in the feedlot. The simulated environment is that of a typical Northeastern Colorado range operation, and forage parameters have been adjusted accordingly. Cold weather has major influences on gains and maintenance requirements during Colorado winters, so an additional subroutine was developed to simulate effects of cold. In order to bet-

Department of Animal Sciences, Colorado State University, Fort Collins, Colorado.

ter account for supplementation on native range during the winter (a common practice in this region), complete ration digestibilities and crude protein contents were not fixed, but were allowed to vary according to animal preference and limits on intake. Calving difficulty was added to the model as a function of cow size and calf birth weight.

The most striking change in the biological program is an increase in the number of traits (genetic potentials) simulated (Table 1). An animal's growth curve is described by three values: Birth weight (BWA), yearling weight (YW) and mature weight (WMA). Milk production (PMA) is the same as in the Texas model. Four genetic potentials for fertility are simulated: age at puberty (AAP), gestation length (GL), maximum probability of initiation of estrus (PEA) and maximum probability of conception given estrus (PCA). Other variables include digestive capacity (DC) -- a factor affecting physical intake limits, mature body composition (FA), and a measure of insulation due to hide and hair (EINS). Some of the new traits are simply variable versions of what had been constants in the original model; calculation of others required considerable programming.

Table 1. Simulated Genetic Potentials.

BWA	--	birth weight
YW	--	yearling weight
WMA	--	mature weight
PMA	--	milk production
AAP	--	age at puberty
GL	--	gestation length
PEA	--	maximum probability of initiation of estrus
PCA	--	maximum probability of conception given estrus
DC	--	digestive capacity
FA	--	mature body composition
EINS	--	external insulation from hide and hair

THE ECONOMIC MODEL

Because our objectives involved economic comparison of genotypes and traits, a program for economic analysis of simulated results was developed using Eastern Colorado ERS data. The economic program attaches directly to the biological model, providing economic analysis of each simulation run. This feature is useful not only because it produces instantaneous economic results, but because it enables us to determine economically optimal levels of supplementation for each genotype, which in turn allows more equitable comparisons of genotypes.

A major constraint of the model is a fixed amount of available range. The number of animals that can be run on a fixed land base varies with genotype of the animal and management policies, and changes in herd size affect the relative contributions of fixed and variable costs to total costs. It was necessary, therefore, to find a method for estimating herd size in different situations. We developed a linear program for this purpose (Table 2). The LP maximizes herd size subject to three sets of constraints: (1) a set of constraints designed to prevent overgrazing by establishing a lower limit on the standing crop of available forage; (2) a set which simulates growth and natural decay of vegetation; and (3) a set which uses forage consumption information from the biological model to calculate changes in total standing crop of forage. The herd size LP is necessarily simplistic, but at least provides an objective way of calculating limits on herd size for different cattle under different management alternatives. Figure 1 represents schematically the relationships among model inputs, the biological model, the herd size LP, and the economic analysis program.

The economic analysis itself is straightforward. Capital assets are assessed ownership costs which include interest, depreciation, taxes, and insurance. Interest on operating expense is calculated using monthly accounting of cash flow. The products considered in the model are live weight at weaning (LWW), empty body weight at slaughter (EBW) and fat-free weight at slaughter (FFW), and economic efficiency is measured by cost per unit of product. With efficiency defined in this way, prices paid for products are of little importance except as they affect cash flow and the relative value of different classes of products, e.g. cull cows versus weaned calves.

Seven economic scenarios are analyzed for each run of the model (Table 3). They represent differences in: the basis for profit determination (the situation of the investor versus that of the rancher who owns his land and cattle and has no alternative use for them),

Table 2. Linear Programming Model for Herd Size
          Determination.

Objective function: Max X_{25} subject to:

Constraint

1. to 12. $X_i > 30$ $i = 1, 12$ Overgrazing constraints

Forage growth and decay equations

13. $X_{13} > .067 (X_1 - 30)$

14. $X_{14} > .071 X_2 - 30)$

15. $X_{15} > .077 (X_3 - 30)$

16. $X_{16} < 4.53$

17. $X_{17} < 27.20$

18. $X_{18} < 36.27$

19. $X_{19} > .316 (X_7 - 30)$

20. $X_{20} > .308 (X_8 - 30)$

21. $X_{21} > .111 (X_9 - 30)$

22. $X_{22} > .063 (X_{10} - 30)$

23. $X_{23} > .167 (X_{11} - 30)$

24. $X_{24} > .200 (X_{12} - 30)$

Total standing crop equations

25. $X_1 < X_{12} - X_{24} - (1.3\ a_{12}/T)\ X_{25}$

26. $X_2 < X_1 - X_{13} - (1.3\ a_1/T)\ X_{25}$

27. $X_3 < X_2 - X_{14} - (1.3\ a_2/T)\ X_{25}$

28. $X_4 < X_3 - X_{15} - (1.3\ a_3/T)\ X_{25}$

29. $X_5 < 30 + X_{16} - (1.3\ a_4/T)\ X_{25}$

30. $X_6 < X_5 + X_{17} - (1.3\ a_5/T)\ X_{25}$

31. $X_7 < X_6 + X_{18} - (1.3\ a_5/T)\ X_{25}$

32. $X_8 < X_7 - X_{19} - (1.3\ a_7/T)\ X_{25}$

33. $X_9 < X_8 - X_{20} - (1.3\ a_8/T)\ X_{25}$

34. $X_{10} < X_9 - X_{21} - (1.3\ a_9/T)\ X_{25}$

35. $X_{11} < X_{10} - X_{22} - (1.3\ a_{10}/T)\ X_{25}$

36. $X_{12} < X_{11} - X_{23} - (1.3\ a_{11}/T)\ X_{25}$

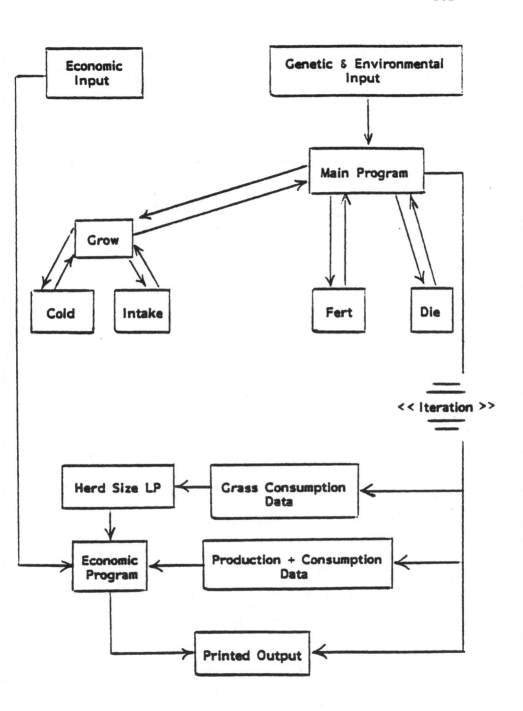

Figure 1. Model Structure.

costs of feeds for the cow herd and feedlot animals, and relative prices paid for products from cull cows versus weanlings or fed animals. In keeping with the overall modeling goal of determining relationships among variables, our approach has been to simulate a number of scenarios simultaneously and then look for trends in the results.

Table 3. Economic Scenarios.

Scenario	Description
1	Standard costs and price relationships, purchased hay, minimal investment
2	Maximal investment
3	Doubled hay cost
4	Doubled processed feed cost
5	Home-grown hay
6	Cow/fed price ratios increased 25%
7	Cow/fed price ratios decreased 25%

COMPUTATIONAL CONSIDERATIONS

The current CSU model runs on a Cyber 205 "supercomputer." The big computer offers both speed and economy; a single run costing approximately $6.00 takes about a minute to produce, depending on demand for computer time. The computer's efficiency allows flexibility in our approach to modeling. We can afford, both in terms of money and time, to "play" with the model, trying different things here and there, fine tuning the feeding of each genotype, and so on. The drawback of using the supercomputer is that the model is not portable without certain modifications.

USES OF THE MODEL TO DATE

The CSU model has been used to compare genotypes for both biological and economic efficiency and to evaluate the relative importance of cattle traits in a rotational crossbreeding system. Cattle types were compared under several management policies within each of the economic scenarios described earlier.

The model has also been used to compare the following management alternatives: varying levels of supplementation, different feedlot rations, varying slaughter endpoints, stocker versus custom feeding programs, heiferette production, and sex controlled production systems.

A SAMPLE OF MODEL RESULTS

A major management decision to be made by producers who have the option of retaining ownership of calves to slaughter is whether to place those calves in the feedlot soon after weaning (weanling system) or to hold them over through their second summer and place them in the feedlot at approximately 18 months of age (stocker system). The simulated effects of changing from the weanling system to the stocker system are shown in Table 4. The stocker system required a reduction in the size of the cow herd, but produced slaughter animals which were older, heavier, and which required less time and feed in the feedlot to reach market condition. The stocker system was generally less biologically efficient than the weanling system, but at least under the conditions simulated, was more economically efficient. What is of particular interest from an animal breeding standpoint is that the smaller genotypes benefitted much more from the stocker system of management than the larger genotypes.

Table 4. Effects of Changing to a Stocker System From a System in Which Weanling Calves Are Custom Fed.

| | | | Slaughter Steers | | | | |
Genotype	Herd size	Sl. age	Time on feed	Sl. wt.	Biol. eff.	Econ. eff.	Net profit
	(%)	(mos)	(mos)	(%)	(%)	(%)	($)
Large cow (1404 lbs)	-15	+6.7	-2.3	+21	-4	+4[a]	+4,598[a]
Medium cow (1179 lbs)	-15	+7.6	-1.4	+32	-1	+5	+7,892
Small cow (955 lbs)	-13	+8.2	-.8	+43	+2	+8	+12,217

[a]Steer slaughter weights above market requirements.

Listed in Tables 5 and 6 are simulated weighting factors for several traits. These represent the changes in overall herd efficiency resulting from independent increases in each trait. The values have been standardized in such a way that they are directly comparable across traits. Such weightings provide a means of measuring the importance of traits in a selection program. The weightings for birth weight in Table 5 are consistently negative, indicating a reduction in efficiency resulting from calving losses associated with larger birth weights. In contrast, weightings for yearling weight are consistently positive and large, showing the importance of rapid early growth. The relatively small weightings for mature weight suggest that in the environment simulated, other measures of growth being held constant, mature cow size has little effect on overall herd efficiency.

Table 5. Weightings Per Genetic Standard Deviation Increase in Each of Three Traits.[a]

| | | Economic Efficiency | | |
Trait	Biological efficiency	Standard costs	Doubled hay cost	Doubled concentrate cost
Birth weight	-.09	-1.58	-3.65	-1.44
Yearling weight	.24	4.76	8.74	4.76
Mature weight	-.01	.48	.57	-.51

[a]Listed values represent changes in efficiency per genetic standard deviation increase in a trait, where biological efficiency is measured as kg TDN/kg EBW and economic efficiency is measured as $/100 kg EBW. Positive values are favorable, indicating increased efficiency.

The weightings for milk production in Table 6 reveal interesting interactions among genotypes, economic scenarios and management systems. Under the weanling management system, increased milk production was biologically inefficient and was economically inefficient when the costs of supplementing the cow herd were high; heavier milking cows required more supplement. When feedlot costs were high, however, increased milk paid for itself; it made economic sense in this case to put as many pounds as possible on calves before they entered the feedlot. Weightings for milk production under the stocker system of management followed the same trend, but were smaller. Milk production was less influential because of the lag between weaning and the time animals were placed in the feedlot.

Table 6. Weightings Per Genetic Standard Deviation Increase in Milk Production.[a]

| | Economic Efficiency | | |
Biological efficiency	Standard costs	Doubled hay cost	Doubled concentrate cost
1. Weanling calves fed out			
-.09	-.17	-1.45	2.79
2. Stocker operation			
-.02	-.18	-.25	.69

[a]Listed values represent changes in efficiency per genetic standard deviation increase in milk production, where biological efficiency is measured as kg TDN/kg EBW and economic efficiency is measured as $/100 kg EBW. Positive values are favorable, indicating increased efficiency.

More interactions are depicted in Table 7, this time among management system (stocker vs weanling), economic scenario ("standard" vs higher ratio between prices paid for cull cows and fed animals), and level of supplementation for the cow herd as reflected in pregnancy rate. Under the weanling system with standard price relationships, the optimal pregnancy rate was only 78%. This value represented the optimal combination of feeding costs and products from both calves and cull cows. (It is important to understand that the model assumes pregnancy testing, culling of open and unsound cows, and retention of a sufficient number of replacements, so that weaning rates need not decline with lower pregnancy rates.) When the price paid for cull cows decreased, the optimal pregnancy rate increased because calves contributed proportionately more to income, favoring the marketing of more heifers and fewer cows. The same pattern was evident under the stocker system of management, only optimal pregnancy rates were higher because the heavier slaughter weights of young stock caused heifers to be more valuable relative to cull cows.

Table 7.   Optimal Pregnancy Rates -- 3 to 10 Year Old Cows.

Management system	Economic scenario	Pregnancy rate
Weanling calves fed out	Standard price relationships	78%
Weanling calves fed out	Cull cow prices decreased 25%	83%
Stocker operation	Standard price relationships	83%
Stocker operation	Cull cow prices decreased 25%	88%

DESIGN AND APPLICATIONS OF A NEW MODEL

We are in the process of developing a new model at CSU. Like the most recent TAMU model, the new model will simulate individual animals using stochastic elements. Unlike the TAMU model, however, our model will support variability in the genetic make-up of individuals. It will be more flexible than our current model in the types of management options possible and may include a more truly interactive animal/forage interface. The new model will be able to simulate successive years with changing inputs. Finally, the economic analysis will be more sophisticated; it will do a better job of determining prices and will include profit as an efficiency measure.

Applications of the model are many. We hope to simulate different combinations of management policies, mating systems, natural environments, cattle types, and economic scenarios. The following is a list of potential applications by category: <u>management policies</u>: calving season, supplementation, pest and disease control, selection and culling practices, retained ownership; <u>mating systems</u>: rotational systems (in space and time), use of heifers, bulls and terminal sires, composites, combination systems; <u>natural environments</u>: forages varying in quality and quantity, different climates; <u>cattle types</u>: different genotypes and combinations of genotypes; <u>economics</u>: varying costs and prices, different market requirements, scales of operation, returns to different economic bases.

FUTURE USE OF BEEF PRODUCTION MODELS

The model described here is and will be a useful research tool for animal scientists and economists. It can serve both to point out relationships among variables and to model specific cattle operations. It is, however, just one of a number of complementary beef cattle models which we see being used in the future. There is a need for both more detailed and simpler models. The detailed models will be strictly biological, simulating a single animal or cow-calf pair. These models should enhance our understanding of basic biological mechanisms.

One simpler model will be what we term a "simulation simulator," a model which produces essentially the same kinds of results as the big bio-economic model, but which is simplified to the greatest degree possible and has modest computational requirements. Developing this model will test our understanding of the relationships uncovered by the larger model and will determine just how far the simplification process can go. If successful, the simulation simulator should be usable by researchers, extension personnel, consultants, and producers. Its equations should be the raw material for spreadsheets and linear programs.

Perhaps the untimate tool for individual producers will be a combination of simulation simulator, spreadsheet, data base for cattle and other records, and farm accounting package. The simulation/spreadsheet will add predictive capability to the historical perspective provided by the data base/accounting package.

REFERENCES

Notter, D.R. 1977. Simulated Efficiency of Beef Pro-
    duction for a Cow-calf Feedlot Management System.
    Ph.D. dissertation. University of Nebraska, Lin-
    coln.

Sanders, J.O., and T.C. Cartwright. 1979a. A General
    Cattle Production Systems Model. Part I: Structure
    of the Model. Agr. Systems. 4:217.

_____. 1979b. A General Cattle Production Sys-
    tems Model. Part II: Procedures Used for Simulat-
    ing Animal Performance. Agr. Systems. 4:289.

X

A Dynamic Model of Beef Cattle Growth and Composition

James W. Oltjen

INTRODUCTION

Simulation models for beef cattle characterize most observed animal growth responses and patterns, but they do not embody the fundamental processes involved in growth. As with growth curves, their use may be limited if extension to new situations is attempted. Therefore, more basic elements must be included in growth models. Postnatal growth generally proceeds by cell division (hyperplasia) and enlargement (hypertrophy). In muscle, nuclear units (DNA) also proliferate allowing increased hypertrophy or protein growth. These processes have been characterized at the tissue and organ level by Baldwin and Black (1979) and Burleigh (1980). By aggregation of these concepts to the whole animal level, Oltjen (1983) developed a dynamic model of post-weaning growth and body composition of the beef steer.

Initially, a submodel of whole body protein growth was developed for the growing rat because only fragmentary DNA and protein synthesis data exist for the bovine (Oltjen, et al., 1985b). Frame size, or mature body weight, affects nucleic acid content, which drives protein synthesis. Accepted equation forms reflecting characteristic rates of rat lean body mass accretion were then fit to cattle (Oltjen, et al., 1985a). Effects of nutrition on DNA and protein synthesis were determined using a large set of steer data containing records of body weight, composition and energy intake. Net synthesis of body fat is calculated as net energy available after accounting for maintenance and protein accretion. Body weight and composition, and mature

Animal Science Department, Oklahoma State University, Stillwater, Oklahoma.

body weight are required as initial inputs; feed intake
and net energy concentrations for maintenance and gain
are input daily and body weight and composition are
simulated.

The objectives of this paper are threefold.
First, a computer model of beef cattle growth is
described. In the next section the results of the
model are evaluated. Finally, the usefulness of the
model to simulate performance of cattle for economic
projection is discussed.

DESCRIPTION OF THE MODEL

Differential equations used in the dynamic deterministic model for growth are

$$\frac{dDNA}{dt} = .00429 * (DNAMX - DNA) * NUT1,$$

$$\frac{dPROT}{dt} = .0472 * DNA^{.73} * NUT2 - .143 * PROT^{.73},$$

where DNA and PROT are empty body DNA (g) and protein
(kg), respectively, DNAMX is DNA at maturity, and t is
time (d). The protein synthesis parameter (.0472) is
adjusted down 4 percent for nonimplanted cattle. NUT1
and NUT2 are scalars incorporating effects of nutrition

$$NUT1 = .-70 + 1.70 * P,$$

$$NUT2 = .83 + .20 * P / (.15 + P),$$

where P is the ratio of metabolizeable energy intake
(MEI, mcal/d) to that required for normal, unrestricted
growth (MEINORM).

$$MEINORM = (.4380 - .2615 * EBW / EBWM) * EBW^{.73},$$

where EBW is empty body weight (kg), and EBWM is EBW at
maturity.

The protein model with the nutritional adjustments
is part of a whole body model consistent with the net
energy format (NRC, 1976). That is, daily maintenance
requirement (MAINT = .084 * EBW$^{.75}$, mcal/d) is supplied
by the net energy for maintenance content of a portion
of feed intake (FI, kg/d), and the remainder of the

feed contains net energy for gain. Net energy for gain is used for protein accretion (DPROT, kg/d), as predicted above, and for net fat synthesis (DFAT, kg/d):

$$DFAT = \frac{(FI - MAINT / NEm) * NEg - DPROT * 5.54}{9.39},$$

where NEm and NEg are net energy concentrations (mcal/kg) in the feed for maintenance and gain, respectively.

In order to use the model as an estimate of initial EBW, DNA and PROT is required, as well as mature EBW (EBWM). DNAMX is directly related to EBWM:

$$DNAMX = 385 * (EBWM / 750),$$

where DNAMX is 395 g and EBWM is 750 kg for an average frame steer. One other frame size adjustment is also used. Because rates of different functions in animals at similar degrees of maturity are proportional to mature weight (EBWM) to the .73 power (Taylor, 1980), rate constants in the model are transformed so that:

$$R'(c') = R(c) * (EBWM' / 750)^{.73},$$

where R(c) is a rate function of c, a component of the animal, and the primed quantities are adjusted for EBWM = EBWM'.

EVALUATION OF THE MODEL

In this paper the model will be compared to commonly used models of cattle growth, the National Research Council (NRC, 1976) and Agricultural Research Council (ARC, 1980) feeding systems, and to the Fox and Black (1984) modification of the NRC. A complete evaluation of the model may be found in the thesis by Oltjen (1983).

Two data sets were used in this evaluation. The first includes data on the growth of 77 distinct groups of steers varying in frame size, initial weight and composition, feeding regime, and growth promotant treatments (Newland, et al., 1979; Coady, et al., 1979; Barber, et al., 1981; Old and Garrett, 1985; Garrett, unpublished data). Subsets of these data had been used to test concepts and estimate parameters in development of the model. In some comparisons only appropriate data were used; nonimplanted cattle were deleted when using the NRC system and implanted cattle were deleted when using the ARC system. A second, independent data set was taken from the USDA Meat Animal Research Center (MARC) cycle one (Smith, et al., 1976) and cycle two (Cundiff, et al., 1981) trials. Evaluations described below were conducted by initializing each model with beginning body weight and composition, and using observed feed intake and feed energy concentrations. Predicted final body weight and composition at the conclusion of the feeding periods are compared with observed data. Body composition for steers in the MARC studies was not measured, so initial composition was set at 10 percent empty body fat.

Residual (the difference between observed and predicted) final body weights are summarized in Figure 1. The mean residual body weight, or bias, is a measure of the accuracy of each system; the standard deviation of the residual (not the standard error of the mean residual) reflects precision. For the first data set, accuracy is similar for both the model (-.4 kg) and the Fox and Black system (-1.5 kg). The NRC tends to underpredict body weight (+14.1 kg); the ARC overpredicts (-23.7 kg). For data appropriate to each system, bias follows the same pattern as before. The model and the NRC are of similar precision, with standard deviations of the residual of 17.6 and 17.3 kg, respectively. The Fox and Black and ARC models are less precise with standard deviations of 34.2 and 28.3 kg, respectively.

For the cycle one and cycle two MARC steers, each system overpredicted the performance of cattle except for the model in cycle one. It also predicted absolute performance more closely than the other systems; the NRC, which has no frame size adjustment, had greatest absolute error of prediction. The best precision was for the model and ARC in cycle one; all systems were similar in cycle two.

Sources of variation in each system were explored by regressing residual against energy intake, initial composition and frame size (Table 1). Each system accounted for variation in energy intake well, with no significant correlation with P. However, a correlation coefficient of .-59 for initial fat content in the ARC system shows that gain is overpredicted for initially fatter cattle. The opposite relationship for the Fox and Black system was observed (r = +.48), possibly due to an overcorrection for rate of gain previous to the feeding period.

The correlations between frame size and residuals are most revealing. In the first data set, larger frame size cattle were underpredicted by the NRC (r = +.54) and overpredicted by the ARC (r = -.70). The effect was confounded by initial composition in this data, with a correlation coefficient of +.35 between initial fat and EBWM. For the MARC data, which were

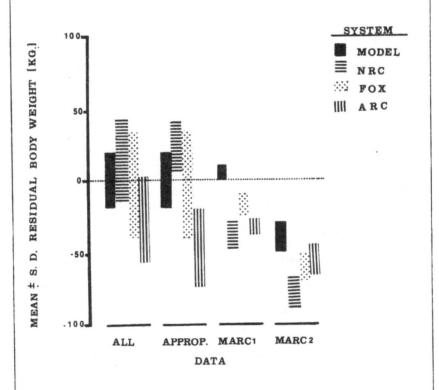

Figure 1.   Residual Final Body Weight and Its
            Standard Deviation (S.D.) in a Com-
            parison of Four Feeding Systems
            (for Data Set Description, See Text)

Table 1. Simple Correlation Coefficients for Relationships with Residual Body Weight.

Item[1]	System:	Correlation coefficient (r)			
		Model	NRC	FOX	ARC
P[2]		-.13	-.14	-.32	+.10
Initial FAT[2]		-.00	-.02	+.48	-.59
EBWN[2]		-.02	+.54	-.08	-.70
EBWN[3]		-.45	+.90	-.68	-.78
EBWN[4]		-.25	+.55	-.43	-.68

[1]P is proportion of normal energy intake, FAT is empty body fat, EBWM is mature empty body weight.

[2]All appropriate data using the first data set for 77 steer groups (see text).

[3]Meat Animal Research Center cycle one steers.

[4]Meat Animal Research Center cycle two steers.

more evenly spaced in regards to EBWM, the NRC clearly underpredicts body weight gains of larger frame cattle and overpredicts those of smaller cattle (r = +.90 and +.55). The FOX and ARC systems, on the other hand, do the opposite, perhaps their frame size adjustments are too large. The model tends towards overcompensation also, but the correlation coefficients (-.25 and -.35) are less than those for the other systems.

Although the model works well for those situations for which it was developed, the system cannot cover all possibilities encountered by growing cattle. It has no provision for heifers or bulls, for inadequate nutrient intake other than energy, for disease states, for harsh environmental conditions, or for feed intake prediction. Its use without regard to these limitations is compromised. However, the present model should be adaptable for extension to additional applications.

ECONOMIC PROJECTION WITH THE MODEL

Development of systems to predict cattle growth at present is entering a new generation. The traditional static systems are being converted to simulation models, and application to a wider range of input conditions is being sought. Fundamental knowledge of biological phenomenon is necessary to characterize animal response to this larger set of inputs, and models founded on such principles will be flexible and useful.

The implications for using a dynamic system model based on such concepts as described above are those associated with increased information for management of modern beef cattle enterprises. Effects of different feeding regimes for cattle of various frame size and condition on steer weight and composition may be simulated. The composition projection is a valuable tool for evaluating different marketing alternatives, particularly for energy level and frame size interactions. Coupled with a management system capable of making projections of cash flow and net returns (e.g. Oltjen, 1985a), the type of cattle and the sequence and duration of rations may be optimized. The value of cattle may be based on their condition and potential for compensatory growth. In short, planning and decision making may be enhanced.

REFERENCES

Agricultural Research Council. 1980. The Nutrient
    Requirements of Ruminant Livestock. Commonwealth
    Agricultural Bureau.

Baldwin, R.L. and J.L. Black. 1979. Simulation of the
    Effects of Nutritional and Physiological Status on
    the Growth of Mammalian Tissues: Description and
    Evaluation of a Computer Program. CSIRO Anim. Res.
    Lab. Tech. Paper No. 6.

Barber, K.A., L.L. Wilson, J.H. Ziegler, P.J. LeVan and
    J.L. Watkins. 1981. "Charolais and Angus Steers
    Slaughtered at Equal Percentages of Mature Cow
    Weight." II. "Empty Body Composition, Energetic
    Efficiency and Comparison of Compositionally Simi-
    lar Weights." J. Animal Science. 53:898.

Burleigh, I.G. 1980. Growth Curves in Muscle Nucleic
    Acid and Protein: Problems of Interpretation at
    the Level of the Muscle Cell. In, Growth in
    Animals. p. 101. T.L.J. Lawrence (Ed.). Butter-
    worths, London.

Coady, M., F.M. Byers, and C.F. Parker. 1979.
    "Protein Requirements for Protein Accretion in
    Hereford X Angus and Charolais Steers." Ohio Anim.
    Res. Dev. Center Beef Res. Rep. Anim. Sci. Series.
    79-1:60.

Cundiff, L.V., R.M. Koch, K.E. Gregory, and G.M. Smith.
    1981. "Characterization of Biological Types of
    Cattle - Cycle II. IV. Postweaning Growth and
    Feed Efficiency of Steers." J. Animal Science.
    53:332.

Fox, D.R., and J.R. Black. 1984. "A System for
    Predicting Body Composition and Performance of
    Growing Cattle." J. Animal Science. 58:725.

National Research Council. 1976. Nutrient Require-
    ments of Beef Cattle. 6th Edition. National
    Academy of Science, Washington, D.C.

Newland, H.W., F.M. Byers, and D.L. Reed. 1979.
    "Response of Three Cattle Types to Two Energy
    Levels and Two Rates of Energy Feeding." Ohio
    Anim. Res. Dev. Center Beef Res. Rep. Anim. Sci.
    Series. 79-1:60.

Old, C.A., and W.N. Garrett. 1985. "Efficiency of Feed Utilization for Protein and Fat Gain in Hereford and Charolais Steers." J. Animal Science. 60:766.

Oltjen, J.W. 1983. A Model of Beef Cattle Growth and Composition. Ph.D. Thesis. University of California, Davis.

Oltjen, J.W., L.G. Burditt, and D.L. Gill. 1985a. An Interactive Microcomputer Program to Simulate Performance of Feedlot Cattle. Oklahoma Agric. Exp. Sta. Software Series.

Oltjen, J.W., A.C. Bywater, and R.L. Baldwin. 1985b. "Simulation of Normal Protein Accretion in Rats." J. Nutrition, 115:45.

Smith, G.M., D.B. Laster, L.V. Cundiff, and K.E. Gregory. 1976. "Characterization of Biological Types of Cattle." 2. "Postweaning Growth and Feed Efficiency of Steers." J. Animal Science. 43:37.

Taylor, St. C.S. 1980. "Genetic Size-Scaling Rules in Animal Growth." Animal Production. 30:161.

XI

Delivering, Updating, and Maintaining Large-Scale Simulation Models: The Role of Professional Agricultural Economists

David H. Laughlin and W. Charles Walden

After visiting with colleagues knowledgeable in the area of biophysical simulation modeling and reviewing the literature concerning delivery of simulation models, we found that there was very little information and few innovative ideas concerning attempts to deliver simulation models and almost nothing concerning updating and maintaining such models. However, the authors hope to explore some questions that will spark thought about methods of delivery of simulation models and the role of professional economists in updating and maintaining them. Servicing the models we build is a very important and often overlooked aspect of most modeling efforts. Unless the information we produce is made available in a usable form to individuals who make resource allocation decisions, our labors will be rather fruitless.

The topic of this volume is the simulation of beef cattle production systems and its use in economic analysis. As evidenced by the excellent papers included which describe three such models and represent a tremendous amount of scientific effort, planning, and experimental resources, there should be no doubt that the biological processes of beef-forage systems can be successfully modeled. Certainly there remains more work to refine and simplify the models, but the basic structure of several models has been cast, and it is now time for model users to enter the picture. We applaud those who have used the products of our biological scientist counterparts in meaningful, useful, often innovative economic studies. Biophysical simulation models have, for several years, been touted as a possible partial replacement for the traditional pro-

Department of Agricultural Economics, Mississippi State University, Mississippe State, MS.

duction function approach to production economics and farm management, and yet few have really dug in to investigate and demonstrate possible applications of these models. Considerable use, validation testing, and incorporation of these models into decision tools in a research mode is a prerequisite to delivery of these models to clientele.

There are a few benchmark studies that have incorporated simulation into economic analysis. Mapp and Eidmann (1976) used simulation to study irrigation timing. More recently Boggess, et al.(1983) used simulation to assess risk-returns of irrigation decisions, and Musser and Tew (1984) expound on uses of biophysical simulation in production economics. In the beef-forage area, the 1970's works of Fox and Black (1977) are the benchmarks for infusion of economics into simulation techniques, or visa versa. In the late 70's several users (Angirasa, et al., 1981; Stokes et al., 1981; Sullivan, 1979) incorporated simulation modeling by Sanders and Cartwright (1979) into economic analyses. About that same time, researchers at Kentucky (Loewer, et al., 1977) were simulating beef production as part of whole farm simulations. More recently Brorson, et al. (1983) developed a stocker cattle simulation model for microcomputers. These studies (even considering the ones we did not find or mention) are relatively few in number when one considers the expertise utilized, time spent, and resources available to model the biophysical aspects of beef-forage production.

There are a few well-founded reasons for this lag or non-use of these models. One reason that has caused such a reluctance to model use is that these models are mathematically and logically complex, and few economists want to invest the time needed to understand the biological processes involved and hence gain confidence in the output of the model.

Model builders have gone to great lengths to model every detail and piece of data that one could imagine so that an adequate trail of events could be studied and analyzed during each simulation run. There is no doubt that this information is useful for validation and some types of experimentation. Most often, however, economists are interested only in the bottom line, e.g., pounds of gain per acre or pounds of forage produced per acre. Certainly these items are present, but the voluminous outputs and complexity of the systems often scares many of us off. I am sure many of you have thought of this problem and no doubt solved it. I know the S-156 modeling effort is now using statistical techniques to reduce data output and analyze simulation results.

Another reason for simulation model non-use is that there has not been a well-defined, complete, and generally accepted theory developed around the use of such models as is available with our neoclassical economic theory which is built upon the production function approach. Our training and professional acceptance induces us to try to incorporate the principles of a theory built around some optimizing mathematics of calculus into a methodology which does not lend itself well to optimization. The point is that we have not yet determined and sufficiently described or demonstrated the role of simulation models as tools that provide information on which to base decisions. There have been a few excellent studies, but not nearly enough and certainly not sufficiently detailed so as to provide decision makers much directly applicable information. Certainly the models themselves have not been put into the hands of producers for their use. If we as researchers have yet to discover and adequately describe how one might use simulation as a basis for decision making, then it may not yet be time to be discussing delivery mechanisms and updating and maintenance responsibilities.

Perhaps this statement is too strong. Maybe we do have at least an elementary handle on simulation modeling and its use in economics and we should investigate these topics. We can simulate the biophysical aspects and at least a few studies have demonstrated some applicability. If we begin to consider how these models might be delivered and who is responsible for maintaining them, then possibly the remainder of the development can skewed to accommodate a broader range of clientele and ultimately make the models more beneficial.

Several questions arise regarding the topic of this chapter. Who are the clientele? Since simulation is purported to be delivered as part of a decision aid, how have decision aids been delivered in the past? What constitutes a large-scale simulation model? What constitutes delivery and how do we measure successful delivery? How will simulation models coupled with economic components likely be delivered? What is maintenance? What is updating? Are the computer model and the economic models developed around them one in the same? In the remainder of this paper we will explore some of these questions and expound on others as they arise.

Clientele

The first major question that must be addressed is "Who are the clientele?" Each of us has an idea about who the clientele are. I suspect that modelers and research users of models should consider agricultural producers as the ultimate beneficiaries of their efforts. However, there are some intermediate clientele groups as well. These groups would include research agricultural economists, extension economists, administrators of agricultural research funds, agricultural students, consulting agriculturalists, financial advisors, and the host of individuals and firms that service and supply inputs to the agricultural sector. These groups may well be the primary clientele for some time because delivery of highly complex simulation models directly to producers may come slower than some of us would like to believe. Until we can nail down some specific analyses that can be performed rather universally, these will remain the primary clientele.

What constitutes a large-scale simulation model? What is a simulation model? The authors consider a biological simulation model to be a set of mathematical-logical statements within a computer which will produce output data that emulate data that would have been produced by a real biological system. Most of us think of a large-scale model as meaning very complex, large models which require large amounts of computer storage space and large amounts of computer memory to operate. This generally means that these models must be run on mainframe type computers. Certainly, all the models of this symposium would be classified as large-scale models under this criterion. But, will these models and subsequent spin-off models be considered large-scale?

Few people envisioned the intensity and rapidity of acceptance of the current microcomputer age. Microcomputer technology of today has transformed many so-called mainframe applications of five years ago and fewer into microcomputer applications of everyday usage. As models themselves become more refined and unnecessary parts desposed of, the models themselves will become smaller and more efficient. Spin-off models or sub-models will be developed and used for specific analyses. There are at present time still some limitations associated with microcomputers, but the largeness problem associated with modeling is fast becoming non-existent. There are other problems associated with delivery of simulation models using microcomputers like appropriate arrangement and design of input and output sequences so as to make them simpler for end users.

Appropriate analysis of results remains another real problem. But these problems center around the sophistication, knowledge level, and analytical skills of the model user rather than the software. There is no doubt that delivery systems involving microcomputers are the wave of the future for a large portion of our computing needs in research, teaching, and extension but more importantly for commercial agriculture. Research and extension had little success in delivery of

computer service to agricultural producers via mainframe type computers during the 60's and 70's. To
limit or target applications solely to mainframe computers would be a serious mistake. Today, many farmers
have their own microcomputer systems and most have
thought about getting one. So, technologically, the
stage is set to deliver to virtually everyone -
researchers, administrators, agribusiness, and commercial producers - virtually any kind of model or simulation technique and accompanying decision aid that we
can develop. The question is, can we develop and deliver decision models sufficiently simple in both input
and output so that the clientele can use them and yet
maintain the integrity of the models? This is the
point at which the clientele must be specifically identified and models and application targeted.

Delivery

What constitutes delivery of a simulation model?
This question and our subsequent discussion will paral-
lel the one about who are the clientele. Delivery of
simulation models will almost certainly take on different forms to different clientele. Delivery of simulation models (or results of simulation models) to commercial agricultural producers will most likely be
filtered through several other so-called clientele
groups in the chain from developers and researchers to
producers. Does the design, execution, and reporting
of results of a contrived, general, representative
situation using simulation models qualify as delivery
of simulation models? Must those results be taken one
step farther and put into the hands of producers before
the model can be considered delivered? Must the model
itself be put into the hands of producers in a form
where they can run the model and analyze results for
themselves before we consider the model delivered?
Certainly delivery of simulation models to clientele
will take on all of these forms at some point in time.

Decisions aids of the past have been delivered in
a variety of ways. Historically, the mode of delivery
has been through the extension service system of agents
and specialists and their educational programs as well
as our undergraduate educational efforts. By and large
the programs and educational efforts have consisted of
demonstrating the various methodologies of budgeting
and marginal analysis applied to different situations,
i.e., things the producer could do with pencil and
paper, using arithmetic or calculator. If a producer
was particularly ambitious we might even try a little
linear programming and/or cash flow analysis. This is
probably an over simplification but the point is to
keep it relatively simple. Today, the situation is

rapidly changing. The microcomputer has provided producers with an exceptionally fast pencil, calculator, and storage place for all kind of methodologies all rolled into one. With some new budget generator software coming out and a little knowledge about marginal analysis, producers (and their bankers) will quickly surpass the educational programs of the past and demand more in the way of data managers, data producers, and data analyzers, i.e., decision aids. This may well be the niche for simulation models and new decision tools that will be forthcoming from the research community. Once again, to reiterate, a major chore for this type of delivery is to couple simulation models with decision models and make imputs and outputs user friendly and unambiguous. As these tools become available and more widely used and as extension users and producers become more familiar and knowledgeable about computer usage, this problem will become less significant. However, this situation is most critical in the short and intermediate term and should not be ignored or shuffled off as the responsibility of someone else. A real educational job is before us.

To date the most common form of delivery of simulation models has been of the strategy development type. Researchers have used simulation models to evaluate specific scenarios for model farm situations and developed generalized strategies to submit to either extension personnel who in turn have contact with producers or researchers working directly with farmers. There have been a few attempts to deliver large-scale models directly to producers (Debertin, et.al., 1981) but most have met with only limited success. At Mississippi State some researchers have actually used a cotton simulation model in a real-time sense to evaluate decisions about spraying for insect control. Researchers are in the fields daily collecting data and periodically updating and realigning the model so that it will make more accurate projections. Preliminary results have suggested that the model be squeezed onto a microcomputer, which they are currently trying to do, but otherwise the modelers are having limited success. Most scientists and model builders, however, do not consider real-time simulation a viable alternative, primarily because environment plays such a large part in responses of biological systems (Dougherty, et. al.,1985).

Another way that simulation models have been delivered directly to producers in a limited manner, initially on mainframe computers and more recently with microcomputer models, is through a type of production game. Weather variables are drawn at random and the user has a chance to make some decisions about spraying, watering, or cutting, depending upon what the decision variables are, and the model registers the production response. The user can follow the production process as though he is really producing the crop and making decisions throughout the year. This type of game gives the user the "feel" for response of the crop to alternative decision strategies. This approach is particularly applicable to classroom type instruction and extension workshop programs. Although this type approach allows the inspection of probable outcomes of many management scenarios, in itself, the technique falls short if one is searching for optimum strategies.

The microcomputer age with all of its advantages of bringing computing capabilities down to the farm level, may well bring with it some disadvantages not yet encountered. A cartoon presented not long ago said "to err is human, but to really screw up you need a computer". If users do not fully appreciate the old computer acronym GIGO (garbage in, garbage out) and maintain considerable diligence in providing correct and appropriate data to decisions models, the microcomputer may well provide a way to make bad decisions on a more timely basis. When simulation models are used in a predictive mode, and then that information is used in decisions models independently by farmers, what are the risks of inaccurate information or making incorrect decisions? Who assumes those risks? The developer of models and decision aids usually disclaims any responsibility. Users on the other hand might not be so understanding after a large monetary loss. Without a doubt the models themselves will be based on some complex logic requiring significant input data and many assumed parameters. More often than not, the users, particularly farmers unskilled in use of models, will assume that since the results come from a computer and the program was developed at the University, they must be correct. We have found it very difficult to get microcomputer users to read instruction manuals let alone understand the principles incorporated within a piece of softwate. This will improve in time but unfortunately it is true today. Who then is responsible for losses that could be attributed to incorrect predictions? Certainly, as models become distributed for microcomputers there will be no way to police their usage and make sure that appropriate input data are used and results are analyzed correctly.

Maintenance and Updating

The title of the paper implies a distinction between maintenance and updating. It suggests the distinction between the two is slight. It is difficult to identify where maintenance ends and updating begins. Perhaps the following definitions will help to separate these two processes. Maintenance is defined as the task of correcting bugs that are discovered in a system during its productive lifetime. Updating is defined as bringing up-to-date. Updating will normally be limited to the data or the selected coefficients necessary for the model's operation.

Heraclitus exclaimed, "There is nothing permanent except change." This statement applies 2,500 years after he made it as we attempt to deal with our computer models and their changing natures. We must keep in mind that the computer is a moron. The computer awaits our use of its power; and, as we implement it in our work, we take on vast responsibilities to monitor, test, maintain, and update it. Given enough time and applications, errors will slowly but surely emerge. These bugs must be located and corrected, and it is the responsibility of the developers to see that these corrections are made. Until we have a model that can, with its associated software and hardware, detect and correct its own mistakes, the author of the model will always be required to accept this task.

The need for updating the model might arise for many reasons. The introduction of new data, methodology, concepts, and technology are but a few of the obvious events that might require the developers to update their computer model. It is difficult to determine the exact moment the update process evolves into a new model. At some point, the developers make a decision that they have a new model that is separate from existing work on the problem. A new model emerges, and maintenance of the new model begins. The fact remains that the developers are the initiators and thereby assume responsibility for their ideas and work. Who else?

One area that model builders can improve upon, whether they be economists or otherwise, is in the structural design. Standards of structural design might be necessary in order to make computer programs more institutional. This approach might be justified on gains in maintenance cost efficiency, portability, and transferability from one group of researchers to the next. If the university invests $50,000 in a computer model under the direction of Dr. Applesnort, it can little afford the loss of that investment because Dr. Applesnort decides one morning he can no longer tolerate the hot weather at the university and moves to Fairbanks, Alaska. Proper design could make maintenance and updating easier and more cost effective.

Concluding Remarks

In summary, we suggest that at least in the near future and possibly intermediate future, simulation models will continue to be used directly by researchers and extension primarily for strategy development and situation evaluation. Although model builders continually try to reduce the input requirements and make models easier to operate, the current state of the art in simulation leaves simulation a very complex analysis tool requiring significant amounts of input data, model assumptions, and biological process understanding that should be recognized before accurate analysis can be made. Very few decision models incorporating simulation have surfaced in a readily deliverable form.

One of the most promising avenues for delivery of simulation and decision aids is via microcimputers. Development and testing of these models for microcomputer delivery is still in its initial stages, and it will be some time until many universally accepted models will be ready for delivery to farmers.

Authors must take responsibility for maintaining and updating their models. However, part of the responsibility as a scientist is to produce a product that can be used, understood, and replicated by his peers. Improvement in initial design of computerized models is clearly possible. Perhaps the long-lived models need additional design structure. The models must be easier to debug so that mundane maintenance can be passed more easily to others. To do this, the scientist needs to learn more about the structural design of programming or to rely more on the professional programmer. Authors of models must accept the fact that such an approach would reduce his importance regarding the use of his models. The institution would become less dependent on his presence in using his particular model. However, the less time the scientists must spend on maintenance and updating, the more time they can spend on improved models or entirely new problem areas.

REFERENCES

Angirasa, A.K., C. Richard Shumway, T.C. Nelson, and
    T.C. Cartwright. 1983. "Integration, Risk, and
    Supply Response: A Simulation and Linear
    Programming Analysis of an East Texas Cow-Calf
    Producer." Southern Journal of Agricultural
    Economics. 13:89-98.

Boggess, W.G., G.D. Lynne, J.W. Lores, and D.P.
    Swaney. 1983. "Risk-Return Assessment of
    Irrigation Decisions in Humid Regions." Southern
    Journal of Agricultural Economics. 15:135-143.

Brorson, Wade E., Odell L. Walker, Gerald W. Horn, and
    Ted R. Nelson. 1983. "A Stocker Cattle Growth
    Simulation Model." Southern Journal of
    Agricultural Economics. 15:155-162.

Cartwright, T.C. "The Use of Modeling to Hasten and
    Enhance the Application of Research Information to
    Beef Production Systems and Feedback to
    Research." Forage-Fed Beef: Production and
    Marketing Alternatives in the South. Bulletin 220
    Southern Cooperative Series, pp.451-457.

Debertin, David L., Charles L. Moore, Sr., Larry D.
    Jones, and Angelos Paqoulatos. 1981. "Impacts on
    Farmers of a Computerized Management Decision
    Making Model." American Journal of Agricultural
    Economics. 63:270-74.

Dougherty, C.T., N. Gay, O.J. Loewer, and E.M. Smith.
    1985. "Overview of Modeling Forage and Beef
    Production." In Simulation of Forage & Beef
    Production in the Southern Region. pp.3-8.
    Southern Cooperative Series Bulletin 308.

Fox, D.G., and J.R. Black. 1977. "System for
    Predicting Performance of Growing & Finishing Beef
    Cattle." In Report of Beef Cattle Feeding
    Research. Michigan Ag. Exp. Sta. Res. Rep. No.
    328, pp. 141-62.

Gallant, John. 1985. "Maintenance Protects Assets."
    Computer World. 11:71-79.

Harsh, Stephen B. 1978. "The Developing Technology of
    Computerized Information Systems." American
    Journal of Agricultural Economics. 60:908-12.

Infanger, Craig L., Lynn W. Robbins, and David L.
    Debertin. 1978. "Interfacing Research and
    Extension in Information Delivery Systems."
    American Journal of Agricultural Economics.
    60:915-20.

Kelly, Derek A. 1983. Documenting Computer
    Application Systems: Concepts and Techniques.
    Petrocelli Publications.

Lacewell, Ronald G., and James M. McGrann. 1982.
    "Research and Extension Issues in Production
    Economics." Southern Journal of Agricultural
    Economics. 14:65-74.

La Due, Eddy L. 1978. "Impacts of Alternative Remote
    Access Computer Systems on Extension Programs".
    American Journal of Agricultural Economics.
    60:135-9.

Lower, Otto J., Jr., Gerald Benock, Nelson Gay, and
    Edward M. Smith. 1977. Production of Beef with
    Minimum Grain and Fossil Energy Imputs. University
    of Kentucky, Lexington, KY.

Mapp, H.P., Jr., and V.R. Eidmann. 1976. "A
    Bioeconomic Simulation Analysis of Regulating
    Groundwater Irrigation." American Journal of
    Agricultural Economics. 58:391-402.

Musser, Wesley N., and Bernard V. Tew. 1984. "Use of
    Biophysical Simulation in Production Economics."
    Southern Journal of Agricultural Economics. 16:77-
    86.

Orwig, Gary W. 1983. Creating Computer Programs for
    Learning. Reston Publishing Company, Reston, VA.

Sanders, J.O., and T.C. Cartwright. 1979. "A General
    Cattle Production Systems Model I: Structure of the
    Model." Agricultural Systems. 4:217-27.

Shneiderman, Ben. 1980. Software Psychology.
    Cambridge, Mass. Winthrop Publishers.

Siess, J.A., and J.B. Braden. 1982. "Online Databases
    Relevant to Agricultural Economics." American
    Journal of Agricultural Economics. 64:761-67.

Stokes, K.W., D.E. Farris, and T.C. Cartwright.
    1981. "Economics of Vertical Integration and Beef
    Production." Paper presented at the Southern
    Agricultural Economics Association meeting in
    Atlanta.

Sullivan, G.M. 1979. "Economics of Improved Management for Transforming the Forage/Livestock System in Tanzania. A Simulation Model." Ph.D. dissertation, Texas A&M University, College Station.

Watson, Vance H., and Chester M. Wells, Jr., Editors. 1985. "Simulation of Forage Beef Production in the Southern Region." Southern Cooperative Series Bulletin 308.

Yourdon, John, and Larry L. Constantine. 1979. Structural Design-Fundamentals of a Discipline of Computer Program and System Design. Englewood Cliffs, N.J. Prentice-Hall.

XII

Administration of Multidisciplinary Research

N.P. Thompson and J.M. Davidson

At some near time it is concernable that all information will be available to all people at all times. We may have a wrist watch with a screen with which we can call up a beef cattle simulator program or on which we will be able to read Moby Dick. These examples may be a bit extreme, but progress is very rapid and we need to think about future possibilities. In this particular context it has been said that within fifteen years, people who do not understand what computers can do will be considered illiterate.

With the prospect of being somewhat elementary in considering multidisciplinary research, let us consider the beginning of a typical career at a land grant school. A review of selection procedures indicates that the department head receives permission to advertise a position, consults with faculty, and prepares a job description. Since what is desired is a good person, the position description is usually somewhat broad rather than focusing too narrowly on a particular area. A candidate is selected and since land grant schools are mission oriented, the individual will investigate state and industry needs and will mold those needs into a program incorporating his or her own background and training. After a period of time, certainly within a year, a faculty member will develop a research program including a research project statement based on the job description and other considerations. Based on the project statement or plan of work, evaluation of progress can be made by administrators. There are several kinds of evaluation. A peer evaluation is a continuing process in a department. There is a sense of whether an individual is performing well or not by listening to

Institute of Food and Agricultural Sciences, University of Florida, Gainesville, Florida.

various faculty comments. Regional or national peer evaluation of faculty based on progress in their subject matter discipline can also be obtained. Another kind of evaluation comes from clientele. This evaluation is important to administrators because clientele will have some comments. They may state that a particular faculty member is the best person we have ever had in that department, they've helped me do this and that... or that person has been in your department and they have not done a thing, or they are supposed to be working in such and such an area but they haven't done anything for me. They've been there four years and I don't know what they're doing.

What we come down to is the term productivity. What is productivity? Productivity can mean different things to different people. One consideration of productivity is cost. It costs the state a certain amount of money for each faculty member. For example, that figure might be $100,000. Most faculty will say I don't make $100,000 but considering bricks and mortar, general support, lights and electricity, technician support, and operating money, total dollars per faculty member is approximatley $100,000. A 30 year career costs $3 million, and the taxpayer sits back and asks what have you done for $3 million. A plant breeder might release a new variety which within a year of its release would generate that amount of money. One or two people might design a computer simulation which is accepted and used which could easily be worth that amount of money in better management, etc.

In an evaluation of the productivity of faculty, refereed and non-refereed publications, popular articles, and technical bulletins are expected as a result of research. Graduate student supervision is considered and if the faculty member is attracting graduate students. Have any creative works been generated, which would include computer programs with appropriate documentation? Have any varieties been released, for example, if the work is in the breeding area? Have outside funds been obtained? University service is also considered. The important consideration is that the faculty are evaluated according to the job description under which they are working, and that all aspects of performance be considered.

Let us focus next on the question of whether interdisciplinary research can be evaluated by the same criteria that we have just discussed or do we have to look for some other criteria? In examining a particular faculty member, it is essential that the person develop a complete program. The junior faculty member in a department cannot be merely a technician or service person. This is an area that must be critically examined when evaluating multidisciplinary efforts. It is essential that each faculty member have an opportunity to have some innovative input into project planning. In a multidisciplinary program each person should be involved in the planning of the work. It is the opinion of some that every person involved in a multidisciplinary project has to do every aspect of the project. This is not necessarily what is meant by multidisciplinary. Each participant should be performing an appropriate portion of the multidisciplinary program. Each should also be involved in publishing the results of the study. In today's environment, multidisciplinary work is absolutely essential in solving our problems. Planning, performing, and publishing: all participating faculty should be involved in each aspect.

The question is, "can a young untenured faculty member afford to become involved in interdisciplinary research?" If so, the administration has to be convinced. The nature of multidisciplinary research has to be understood. Times are changing. Land grant institutions may have become very discipline oriented or commodity oriented and may not always have tried to work together on particular problems. Certainly the Dean or Director has to be convinced that multidisciplinary research is necessary and productive and has to be willing to evaluate people in that context.

The department chairperson is a key person and has the responsibility of guiding young faculty members. He must have a philosophy wherein he is going to support interdisciplinary research and cannot treat young faculty members on a sink or swim philosophy. Rather the department chairperson's job as an administrator is to nurture, bring along, counsel, help young faculty to get his/her research underway and to see that adequate funding is provided. When a person does not qualify for tenure and promotion, some of the guilt has to be placed on the department chairperson. This can be particularly true in multidisciplinary research. The department chairperson should insure that the faculty member is involved in planning research and is not just involved in a service capacity. The time committment of the faculty member should be agreed upon ahead of time. In addition, each young faculty member should have at least one project on which he/she is the principal investigator and not only serving on three of four projects on which somebody else is the principal investigator. Publication policy should be agreed upon ahead of time. Generally the person who does the most work is the senior author; however, it is essential that everyone involved in a multidisciplinary project have an opportunity to be the senior author on some aspect of the work. There may be several facets to a multidisciplinary project with a number of papers being written without redundancy, each speaking to different audiences. It is preferable that a young faculty member not spend all his/her time in multidisciplinary research. Time must be given to self-improvement in a discipline perhaps related to the multidisciplinary project, but where personal visibility can be afforded.

In many of the states, including Florida, there has been a structure for multidisciplinary research. Florida has 20 agricultural research centers throughout the state. These centers were established on interdisciplinary principles. For example, a commodity approach might involve several disciplines such as soils, production, pest management, breeding, postharvest handling, marketing, nutrition, etc. Each faculty member is encouraged to emphasize individual disciplines while involved in cooperative effort.

The integration of research programs with extension activities is an increasingly important aspect of today's agriculture. The "I'm research - you're extension" philsophy, with little personal interaction, is obsolete. This does not mean that there are not individual responsibilities according to job description. The break between research and extension is, and perhaps should be, difficult to define. What must be avoided is strict compartmentalization which eventually results in incomplete dissemination and adoption of project accomplishments. To this end the extension specialist should be involved as early as practicable in research projects. The joint extension-research appointment is a useful mechanism for accommodating useful information transfer.

Interdisciplinary work is more difficult to program than individuals efforts. This fact must have broad understanding and acceptance. However, more difficult, the rewards to the various interested groups, clinetele, and disciplines will be commensurate.

Graduate students should be involved in interdisciplinary projects. A tremendous potential exists for increasing the breadth of education in addition to satisfaction of in-depth requirements of particular disciplines. In the early reconnaissance years of scientific inquiry and accomplishment, the promise and reward of interdisciplinary association is of great value in developing the "complete" faculty member. The thesis or dissertation will have depth in a particular discipline but will also profit by embellishment with wisdom from wider resource bases.

The administration of interdisciplinary programs can be accomplished in several ways. The department with the major contribution can have its chairperson administratively responsible for the program. He/she must be overtly committed to interdisciplinary work in order to quell any notions of favoritism toward his/her discipline. Funding can be provided to the project through this department with other departments receiving allocations via interdepartmental transfers. A modification of this administrative arrangement can be a committee of chairmen with involvement with one serving as chairman of the group. Another administrative mechanism provides for leadership from the Experiment Station office with faculty receiving support earmarked for the project directly to their respective departments. Where large programs of high priority and visibility are contemplated, a major center can be established with the Center Director receiving a total project funding allocation which is then distributed to the faculty through the several departments. As previously noted, interdisciplinary work cannot only be more difficult to perform but also to administer. Nothing can destroy on-going camaraderie in pursuing program objectives more than distrust and misunderstanding in the disbursement of support monies. It is highly desirable that all parties have the opportunity to be fully informed in regard to the planned and actual disbursement of project support dollars. Interdisciplinary research does not differ significantly from other research endeavors from this aspect.

Index